Cosmic Analogies

How Natural Systems Emulate the Universe

Other Title by Valerio Faraoni

Alpine Physics
Science in the Mountain Environment
ISBN: 978-981-3274-20-4

Cosmic Analogies

How Natural Systems Emulate the Universe

Valerio Faraoni
Bishop's University, Canada

 **World Scientific**

NEW JERSEY · LONDON · SINGAPORE · BEIJING · SHANGHAI · HONG KONG · TAIPEI · CHENNAI · TOKYO

Published by

World Scientific Publishing Europe Ltd.

57 Shelton Street, Covent Garden, London WC2H 9HE

Head office: 5 Toh Tuck Link, Singapore 596224

USA office: 27 Warren Street, Suite 401-402, Hackensack, NJ 07601

Library of Congress Control Number: 2022032899

British Library Cataloguing-in-Publication Data
A catalogue record for this book is available from the British Library.

COSMIC ANALOGIES
How Natural Systems Emulate the Universe

ISBN 978-1-80061-342-3 (hardcover)
ISBN 978-1-80061-343-0 (ebook for institutions)
ISBN 978-1-80061-344-7 (ebook for individuals)

For any available supplementary material, please visit
https://www.worldscientific.com/worldscibooks/10.1142/Q0396#t=suppl

Typeset by Stallion Press
Email: enquiries@stallionpress.com

To Louine and Donovan

Preface

According to the Merriam–Webster dictionary, an analogy is "a comparison of two otherwise unlike things based on resemblance of a particular aspect" or the "inference that if two or more things agree with one another in some respects they will probably agree in others" [282]. Quantitative sciences use the common language of mathematics and an analogy can then take the form of an equation, or set of equations, being used to describe very different phenomena or physical systems. A mathematical analogy consists of the same formal mathematics expressing the features, or the dynamics, of systems that are very different in nature but of similar structure, form, or dynamics. The fact that these systems are governed by the same set of equations means that they are structured or function in the same way, albeit in very different contexts. This is sometimes expected but other times it looks remarkable.

We often use analogy as a didactic strategy, especially in the sciences (*e.g.*, [12,63,231,308]). For example, heat conduction is analogous to Ohm's law for a current in a linear resistor, which is the basis for the "R–value" of commercially available insulating panels. In another example, a linear electric circuit containing a resistor, a capacitor, and an inductor is shown to be related with a harmonic oscillator, a particle subject to friction and the restoring force of a spring and endowed with inertia (*e.g.*, [183, 304]). The two systems have very different nature but the concepts that describe them have a generality that transcends them. Analogy is also used for discovery: in mathematics, Polya places it in the category of "plausible" reasoning [333]. It is through analogy with Buckminster's geodesic domes in architecture that the molecular structure of fullerene was explained [247] (see the description of this discovery in Ref. [4]). Quantum chromodynamics was modelled on the existing theory of quantum electrodynamics. Analogies in atomic and subatomic physics are discussed in Refs. [306,441]. Exploring the corners where an analogy breaks down may disclose new phenomena.

It has been remarked that analogies do not really teach anything new, do not develop new ideas because they simply duplicate ideas that already exist, and that they use an existing cognitive structure to understand new systems without generating new concepts. Even as pedagogical tools, they are believed to have many limitations. In defence of analogies, the criticisms are only partially true and they should be taken with a grain of salt. For example, in the analogy between electric circuits and oscillators, the fact that no new knowledge is developed would be true only if we already had total mastery of the concepts of inertia, dissipation, and forces in an abstract way in both systems, but these concepts are learned bit by bit. First, they are learned in mechanics and then recognized to apply to circuits, but in a very different form. In fact, the physical quantities expressing them are defined in very different ways in these two kinds of systems. The analogy helps recognizing that electrical systems behave, in some abstract way, as mechanical systems. Physical intuition hints at that, but it is the mathematics that makes this analogy precise by showing that the same equations rule the two systems, except that the physical meaning of the symbols change. The fact that these systems "work in the same way" would often be missed without a formal analogy. In this sense, analogies do help us discovering something. When intuition fails, an analogy can guide us. For example, before the advent of computers, analogue electrical circuits were built to simulate non–linear systems ruled by equations too complicated to be solved analytically. Building analogue circuits became an art and a science in itself (*e.g.*, [304]), now relegated to the sands of time by powerful computers. These complicated analogue circuits, however, make the point that one can learn something new from analogies.

Other examples in physics show that, even though it is clear that certain systems are analogous because they obey the same mathematics, there is still something to learn from analogies. In the early days of quantum mechanics, the realization that particles can behave as waves, and the application of wave mechanics to them, paved the way for progress. It is true that this is hardly an analogy because quantum particles are really matter waves and wave mechanics, with its general methods, covers quantum phenomena but it is far–reaching to push the analogy and realize, for example, that classical mechanics is just the geometric optics limit of quantum mechanics for small wavelengths [183]. These more comprehensive points of view help deepening the understanding, showing that it may be worth pursuing analogies.

In the past two decades, the field of analogue gravity has emerged, in which physical systems are designed, built, and analyzed in the laboratory to mimic real–life systems of very different nature that cannot be constructed, such as huge black holes, wormholes, or the universe itself. Analogue gravity explores indirectly phenomena that are undetectable or out of reach even in principle, such as quantum field theory in curved space, Hawking radiation from black holes, superradiance, or expanding universes. The analogue systems simulating these exotic phenomena comprise fluid–mechanical flows and vortices [41, 167, 321, 324, 325, 368, 374, 405, 407, 408, 423, 425, 442], Bose–Einstein condensates and other condensed matter systems [19, 20, 35, 58, 84, 121, 124, 159, 160, 166, 175, 229, 245, 259, 289, 335, 429–432, 434], optical systems [86, 87, 336, 369, 409], and a variety of other systems including dielectrics, plasmons, granular matter, and possibly graphene and liquid crystals [164, 285, 300, 341, 352, 367, 376]. The essence of these analogue systems is that waves propagating in them obey equations in which an effective metric tensor appears that mimics a geometry of some relativistic spacetime of interest. Although this effective metric is not dynamical, very rich physics of test fields and particles on these fixed background geometries can be explored. Analogue gravity has developed into a mature field, where several fundamental results have been obtained, and has generated a category of scientists who, by necessity, must be proficient in both theory and experiment. Physics had not seen such figures since the times of Enrico Fermi.

In this book we do not discuss analogue gravity in this sense, but we focus on a different class of analogies. Many formal analogies can be found between different areas of physics, but we will not restrict to physics. We explore formal analogies between the universe, as described by general–relativistic (or Friedmann–Lemaître–Robertson–Walker) cosmology and other systems, mostly natural systems studied in the earth sciences. It is not widely known that analogies exist between cosmology and the equilibrium beach profiles studied by oceanography, the glaciers and ice caps that are the subject of glaciology, the distribution of aftershocks following a main earthquake shock, *etc.* We also discuss formal analogies between cosmology and the physics of cooling and heating (which, in turn, provides an analogy between certain universes and the freezing of bodies of water in winter), and between other types of possible universes and gravitational systems. There is also a well–known analogy between a Newtonian ball of expanding material and the Friedmann equation of relativistic cosmology, which is used (and sometimes abused) to introduce relativistic cosmology. This analogy can be pushed a little and one can analyze the corresponding general–relativistic problem of a ball expanding in empty space.

In some cases, the analogy with gravity helps clarifying contentious issues and clearing problems on the other side of the analogy. Bearing in mind that there is a century of literature on general–relativistic cosmology, it is intuitive that this knowledge could, in some cases, contribute to younger fields of research or fields that develop more slowly because the scientific communities involved are smaller than the cosmology community (this is the case, for example, of the scientific community studying coastal morphology and beach profiles in oceanography). Another aspect is that, in the earth sciences, the emphasis is on collecting field data and processing them rather than on theoretical modelling and little or no effort goes into looking for universal laws. As a consequence, general theoretical methods spanning other sciences are usually not valued nor looked for. In these situations, analogies may help developing the understanding of certain phenomena. Taking pragmatically the dictionary's characterization that "if two or more things agree with one another in some respects they will probably agree in others", it is useful to look for these other common aspects that are not obvious and extend, or complete, the analogy.

The reason why cosmology has so many analogues can be traced to the fact that the Friedmann equation, a first order differential equation which constrains the dynamics of cosmic models in Einstein gravity, resembles the energy conservation equation for a point particle moving in one dimension under the action of a conservative force, and already lends itself to a mechanical analogy [312, 328, 342, 396, 410]. The fact that many physical systems exhibit analogies with one–dimensional motions makes it easier to understand why analogies with cosmology are more common than expected. In the earth sciences, however, these analogies with point particle motion are missed, while they would help classifying analytical solutions of the relevant equations and possibly enlarging the catalogue of known solutions. We take advantage of the opportunity to discuss also these mechanical analogies when we encounter them.

Even though the analogy between the Friedmann equation of cosmology and the energy integral for one–dimensional mechanical motion explains many of the possible analogies, some of them still come as a surprise and those are the most entertaining to study, although they may not always be the most useful.

Another aspect should be emphasized: the analogies developed for one–

dimensional systems usually involve a Lagrangian and a Hamiltonian description and a variational principle enforcing the physics involved. In the applied sciences and in the earth sciences such variational principles, although present, are less frequent than in theoretical physics and the analogies help focussing on them. In particular, analogies offer the chance of solving the inverse variational problem for ordinary differential equations ruling natural phenomena, for which Lagrangians and Hamiltonians are usually not discussed. These variational principles are useful when discussing maximum versus minimum entropy production in the earth sciences—a hotly debated subject in climate science. Sometimes the debate can be settled by examining the second variation of the action integral, as is done here for equilibrium beach profiles and for glacial valley profiles.

I had fun working out the analogies presented in this book and learning about areas of science that are not mine. I hope that the readers will have as much fun as I did.

Valerio Faraoni

List of Notations

m, M	mass
V	volume
A	area
g	acceleration of gravity
G	gravitational constant
γ	surface tension
h, z	elevation or thickness
r, R	radius
t	time
$\vec{x}$	position
$\vec{v}$	velocity
$\vec{a}$	acceleration
$\vec{F}$	force
$\vec{N}$	normal force
τ, σ	stress
E	energy
$\vec{p}$	momentum
T	kinetic energy
V	potential energy
L	Lagrangian
$\mathcal{H}$	Hamiltonian
$\pi_x = \partial L / \partial \dot{x}$	momentum canonically conjugated to the variable x
Q	heat energy
T	temperature
C	heat capacity

μ	friction coefficient
$\vec{\nabla}$	gradient
∇^2	Laplacian
K	curvature index of FLRW spaces
κ	normalized curvature index
η	conformal time
$a(t)$	scale factor of a FLRW universe
$H(t)$	Hubble function
$g_{\mu\nu}$	metric tensor
∇_μ	covariant derivative operator
$\Gamma^\mu_{\alpha\beta}$	Christoffel symbols
$\epsilon_{\alpha\beta\gamma\delta}$	Levi–Civita symbol
$\Box \equiv g^{\mu\nu}\nabla_\mu\nabla_\nu$	d'Alembert operator
$T_{\mu\nu}$	energy–momentum tensor
Λ	cosmological constant
$\mathcal{R}_{\mu\nu\rho\sigma}$	Riemann (curvature) tensor
$\mathcal{R}_{\mu\nu}$	Ricci tensor
$\mathcal{R}$	Ricci scalar
$\mathcal{C}_{\mu\nu\rho\sigma}$	Weyl tensor
$E_{\mu\nu}$	electric part of the Weyl tensor
$H_{\mu\nu}$	magnetic part of the Weyl tensor
ρ	energy density
P	pressure
$\equiv$	equal by definition
$\simeq$	approximately equal

Acknowledgments

I am grateful to my collaborators and/or former students Farah Atieh, Adriana Cardini, Steve Dussault, and Andrea Giusti, with whom I worked out part of the material contained in this book; then, to Andrzej Borowiec, Sante Carloni, Justin Feng, Antonino Flachi, Uwe Fischer, Michael Hobson, Roberto Garra, Sergei Kopeikin, Frank Morgan, Germain Rousseaux, Aleksander Simonič, Vladimir Smirnov, Marek Szydłowski, Vincenzo Vitagliano, and Brad Willms for discussions, comments, or for pointing out typographical errors or useful references; and also to the anonymous referees of several journals for valuable comments on articles that, in some form, made their way into this book or influenced it. Finally, I am grateful to Louine Niwa, who prepared the photographs for inclusion in this book, and to Dr. Elena Nash, World Scientific Editor, for her encouragement in bringing this book project to completion.

Contents

List of Figures

Chapter 1

Relativistic cosmology: a primer

Growth comes through analogy;
through seeing how things
connect, rather than only seeing
how they might be different.

Albert Einstein

For the convenience of the reader, and with the purpose of establishing the notation, in this first chapter we review Einstein's theory of general relativity and its description of cosmology. These subjects are huge and, understandably, they cannot be learned in a single chapter or book. We emphasize the basic ideas, with some bias towards the material needed in the rest of this book, and we refer the reader to standard textbooks in general relativity [82, 120, 202, 435] and cosmology [244, 260, 261, 328, 440] for proper treatments.

1.1   General relativity

The physical foundation of general relativity is the Equivalence Principle expressing and generalizing the universality of free fall discovered by Galileo Galilei. This basic experimental fact motivates the use of curved manifolds to describe spacetime, which has the consequence that differential geometry is the language of general relativity.

1.1.1   *Equivalence Principle*

The Equivalence Principle (EP) generalizes Galilei's observation that bodies subject only to gravity fall with the same acceleration. There are three formulations of the Equivalence Principle [443, 444]:

- *Weak Equivalence Principle* (WEP): In the absence of other forces, all bodies fall with the same acceleration in a gravitational field, regardless of their mass, internal structure, or chemical composition. This is Galilei's experimental finding that appears as a coincidence in Newtonian gravity: it is a statement about mechanics.
- *Einstein Equivalence Principle* (EEP): All laws of (non–gravitational) physics in a freely falling frame must reduce to the laws of special relativity. This is a much stronger statement than WEP since it is not limited to mechanics but applies to all physical laws except gravitation itself. For example, electrodynamics in a small freely falling laboratory must reduce to the electrodynamics known in special relativity. The EEP implies local Lorentz invariance. The EEP is a local statement in the sense that it applies to regions of spacetime smaller than its curvature radius, so that the effects of gravity are locally negligible.
- The *Strong Equivalence Principle* (SEP) consists of three parts:
  - WEP is valid for self–gravitating bodies as well as for test bodies (GWEP);
  - the outcome of any local test experiment is independent of the velocity of the (freely falling) apparatus;
  - the outcome of any local test experiment is independent of where and when in the universe it is performed (local position invariance).

  SEP is a stronger statement than EEP since it includes the self–gravity of the body as a contribution to its mass. To test SEP, one must consider a freely falling laboratory small enough to be able to neglect tidal forces from the external gravitational field, but large enough to test its self–gravity (*e.g.*, a Cavendish experiment, or the dynamics of a neutron star).

The Schiff conjecture [366] states that WEP implies EEP but is unproven due to the difficulty of performing precise experiments with self–gravitating bodies. However, SEP implies WEP.

The reason to distinguish between WEP, EEP, and SEP is that there are many theories of gravity alternative to general relativity [443, 444]. Any metric theory of gravity, *i.e.*, one in which gravitation is described as a geometric phenomenon and freely falling particles move on geodesics, satisfies WEP. Tests of the WEP such as the Eötvos experiment and tests of the gravitational shift do not test general relativity but only the metric foundations of relativistic gravity [443, 444]. They are shared by general relativity and by a number of other theories. To put it in other words, Eötvos experiments and tests of the gravitational shift do not discriminate between general relativity and other metric theories of gravity [443, 444].

Einstein's basic idea in describing gravity embodies the Equivalence Principle: since all bodies fall with the same acceleration when subject only to a gravitational field, gravity is a universal phenomenon and can be described by the geometry of spacetime instead of being an accident tied to specific bodies and forces acting on them. Einstein already knew from his 1905 theory of special relativity that a description of physics must be based on a four–dimensional continuum incorporating space and time in one entity, *i.e.*, spacetime. The arena of special relativity, in the absence of gravity, is the four–dimensional flat Minkowski spacetime [136, 250, 342]. When gravity is included in the picture, it is described as the curvature of spacetime. Then the latter must be a curved manifold, described by differential geometry, therefore the language of general relativity (and of all relativistic theories of gravity) is the geometry of curved spaces: differential geometry.

1.1.2 *Differential geometry*

We discuss briefly and without proof the tools of differential geometry needed in the following chapters, referring the reader to [82, 120, 202, 435] for more detailed explanations or proofs.

1.1.2.1 *Manifolds and coordinates*

A curved manifold is a continuum, a set of points which is also a topological space and a Hausdorff space [435]. Its points can be put in one–to–one correspondence with those of the space[1] $\mathbb{R}^4$ by coordinates. Coordinate charts only cover regions of the spacetime manifold, not the entire manifold (unless this is the flat Minkowski spacetime of special relativity), and they can be joined smoothly to each other in regions where they overlap.

One of the main principles of differential geometry, but also of three–dimensional pre–relativistic physics, is that all fundamental laws must be independent of the coordinate system. Coordinates are labels put by humans on space and time and physics should not be tied to them. The

---

[1]We only consider four–dimensional manifolds describing one time and three space dimensions.

Principle of General Covariance states that all laws of physics must be expressed in a form that is the same in all coordinate systems. In differential geometry, this means that these laws must be expressed as tensor equations so that, when a coordinate transformation is made, the left and the right hand sides of the equation transform according to the same law and the form of the equation is preserved in the new coordinate system.

Let us discuss coordinate transformations, which are then used to define tensors. The physically relevant case is $n = 4$, but these definitions can be given for spacetime dimension n.

Let $\{x^\mu\}_{\mu=1,2,\ldots,n}$ and $\left\{x^{\mu'}\right\}_{\mu'=1,2,\ldots,n}$ be two coordinate systems on an open region of the spacetime manifold and consider the coordinate change

$$x^\alpha \longrightarrow x^{\alpha'} = x^{\alpha'}\left(x^\beta\right), \qquad \alpha, \alpha', \beta = 1, 2, \ldots, n. \qquad (1.1)$$

The transformation matrix is the $n \times n$ matrix

$$\left(\frac{\partial x^{\alpha'}}{\partial x^\beta}\right) = \begin{pmatrix} \frac{\partial x^{1'}}{\partial x^1} & \frac{\partial x^{1'}}{\partial x^2} & \cdots & \frac{\partial x^{1'}}{\partial x^n} \\ \frac{\partial x^{2'}}{\partial x^1} & \frac{\partial x^{2'}}{\partial x^2} & \cdots & \frac{\partial x^{2'}}{\partial x^n} \\ \vdots & \vdots & \vdots & \vdots \\ \frac{\partial x^{n'}}{\partial x^1} & \frac{\partial x^{n'}}{\partial x^2} & \cdots & \frac{\partial x^{n'}}{\partial x^n} \end{pmatrix} \qquad (1.2)$$

and its determinant (Jacobian of the transformation)

$$J \equiv \mathrm{Det}\left(\frac{\partial x^{\alpha'}}{\partial x^\beta}\right) \qquad (1.3)$$

must be non–vanishing in order for the transformation to be an an admissible coordinate transformation, *i.e.*, one–to–one, smooth, and with smooth inverse. The inverse transformation is

$$x^{\alpha'} \longrightarrow x^\alpha = x^\alpha\left(x^{\mu'}\right), \qquad \alpha, \alpha', \mu' = 1, 2, \ldots, n \qquad (1.4)$$

and has Jacobian determinant

$$\mathrm{Det}\left(\frac{\partial x^\alpha}{\partial x^{\beta'}}\right) = J^{-1}. \qquad (1.5)$$

We adopt the Einstein summation convention according to which the summation symbol $\sum$ is omitted when an index is repeated, appearing once in a low and once in a high position. For example, $A_\alpha B^\alpha$ denotes $\sum_{\alpha=1}^{n} A_\alpha B^\alpha$.

1.1.2.2 *Tensors*

A contravariant vector (or contravariant tensor of rank 1) is a set of n components A^α in a specified coordinate system $\{x^\mu\}_{\mu=1,2,\,\dots\,,n}$ that transform according to

$$A^{\alpha'} = \frac{\partial x^{\alpha'}}{\partial x^\beta} A^\beta \tag{1.6}$$

under the coordinate change $x^\mu \longrightarrow x^{\mu'}$.

The concept of vector is defined geometrically, *i.e.*, in a way independent of the coordinate system. On the contrary, its components A^α change with the coordinate system and are a representation of that vector on a certain basis.

A covariant vector (or dual vector, one–form, or covariant tensor of rank 1) is a set of n components A_α (note the position of the index) that transform according to

$$A_{\alpha'} = \frac{\partial x^\beta}{\partial x^{\alpha'}} A_\beta \tag{1.7}$$

under the coordinate change $x^\mu \longrightarrow x^{\mu'}$.

Passing to quantities with two or more indices, we define tensors.

A contravariant tensor of rank 2 is a set of n^2 components $T^{\alpha\beta}$ that transform according to

$$T^{\alpha'\beta'} = \frac{\partial x^{\alpha'}}{\partial x^\rho} \frac{\partial x^{\beta'}}{\partial x^\sigma} T^{\rho\sigma} \tag{1.8}$$

under the coordinate change $x^\mu \longrightarrow x^{\mu'}$.

A covariant tensor of rank 2 is a set of n^2 components $T_{\alpha\beta}$ that transform according to

$$T_{\alpha'\beta'} = \frac{\partial x^\rho}{\partial x^{\alpha'}} \frac{\partial x^\sigma}{\partial x^{\beta'}} T_{\rho\sigma} \tag{1.9}$$

under the transformation $x^\mu \longrightarrow x^{\mu'}$. Two–index tensors can be represented by matrices.

The second derivative of a scalar function $\dfrac{\partial^2 f}{\partial x^\alpha \partial x^\beta}$ is not a covariant two–index tensor. We can easily compute its transformation property:

$$\frac{\partial^2 f}{\partial x^{\alpha'} \partial x^{\beta'}} = \frac{\partial}{\partial x^{\beta'}} \left(\frac{\partial f}{\partial x^{\alpha'}} \right) = \frac{\partial}{\partial x^{\beta'}} \left(\frac{\partial f}{\partial x^{\mu}} \frac{\partial x^{\mu}}{\partial x^{\alpha'}} \right)$$

$$= \frac{\partial^2 f}{\partial x^{\beta'} \partial x^{\mu}} \frac{\partial x^{\mu}}{\partial x^{\alpha'}} + \frac{\partial f}{\partial x^{\mu}} \frac{\partial^2 x^{\mu}}{\partial x^{\alpha'} \partial x^{\beta'}}$$

$$= \frac{\partial^2 f}{\partial x^{\mu} \partial x^{\nu}} \frac{\partial x^{\nu}}{\partial x^{\beta'}} \frac{\partial x^{\mu}}{\partial x^{\alpha'}} + \frac{\partial f}{\partial x^{\mu}} \frac{\partial^2 x^{\mu}}{\partial x^{\alpha'} \partial x^{\beta'}} . \tag{1.10}$$

Unless $x^{\mu} \longrightarrow x^{\mu'}$ is a linear transformation, it is $\dfrac{\partial^2 x^{\mu}}{\partial x^{\alpha'} \partial x^{\beta'}} \neq 0$ and

$$\frac{\partial^2 f}{\partial x^{\alpha'} \partial x^{\beta'}} \neq \frac{\partial x^{\mu}}{\partial x^{\alpha'}} \frac{\partial x^{\nu}}{\partial x^{\beta'}} \frac{\partial^2 f}{\partial x^{\mu} \partial x^{\nu}} ; \tag{1.11}$$

hence, not all objects with indices are tensors unless they transform as such under coordinate changes.

Generalizing to an arbitrary number of indices, a contravariant tensor of rank k is a set of n^k components $T^{\alpha_1 \alpha_2 \cdots \alpha_k}$ that transform according to

$$T^{\alpha'_1 \alpha'_2 \cdots \alpha'_k} = \frac{\partial x^{\alpha_1'}}{\partial x^{\beta_1}} \frac{\partial x^{\alpha_2'}}{\partial x^{\beta_2}} \cdots \frac{\partial x^{\alpha_k'}}{\partial x^{\beta_k}} T^{\beta_1 \beta_2 \cdots \beta_k} \tag{1.12}$$

under the coordinate change $x^{\mu} \longrightarrow x^{\mu'}$.

A covariant tensor of rank k is a set of n^k components $T_{\alpha_1 \alpha_2 \cdots \alpha_k}$ that transform according to

$$T_{\alpha'_1 \alpha'_2 \cdots \alpha'_k} = \frac{\partial x^{\beta_1}}{\partial x^{\alpha_1'}} \frac{\partial x^{\beta_2}}{\partial x^{\alpha_2'}} \cdots \frac{\partial x^{\beta_k}}{\partial x^{\alpha_k'}} T_{\beta_1 \beta_2 \cdots \beta_k} \tag{1.13}$$

under the coordinate change $x^{\mu} \longrightarrow x^{\mu'}$.

Finally, a mixed tensor of contravariant rank ℓ and covariant rank m is a set of $n^{\ell+m}$ components $T^{\alpha_1 \alpha_2 \cdots \alpha_\ell}{}_{\beta_1 \beta_2 \cdots \beta_m}$ that transform according to

$$T^{\alpha'_1 \alpha'_2 \cdots \alpha'_\ell}{}_{\beta'_1 \beta'_2 \cdots \beta'_m}$$

$$= \frac{\partial x^{\alpha'_1}}{\partial x^{\rho_1}} \frac{\partial x^{\alpha'_2}}{\partial x^{\rho_2}} \cdots \frac{\partial x^{\alpha'_\ell}}{\partial x^{\rho_\ell}} \frac{\partial x^{\sigma_1}}{\partial x^{\beta'_1}} \frac{\partial x^{\sigma_2}}{\partial x^{\beta'_2}} \cdots \frac{\partial x^{\sigma_m}}{\partial x^{\beta'_m}} T^{\rho_1 \rho_2 \cdots \rho_\ell}{}_{\sigma_1 \sigma_2 \cdots \sigma_m} \tag{1.14}$$

under the coordinate change $x^{\mu} \longrightarrow x^{\mu'}$.

For example, the three–index tensor $T^{\alpha\beta}{}_{\gamma}$ transforms according to

$$T^{\alpha' \beta'}{}_{\gamma'} = \frac{\partial x^{\alpha'}}{\partial x^{\mu}} \frac{\partial x^{\beta'}}{\partial x^{\nu}} \frac{\partial x^{\tau}}{\partial x^{\gamma'}} T^{\mu\nu}{}_{\tau} . \tag{1.15}$$

A tensor equation

$$A^{\alpha_1 \alpha_2 \cdots \alpha_\ell}{}_{\beta_1 \beta_2 \cdots \beta_k} = B^{\alpha_1 \alpha_2 \cdots \alpha_\ell}{}_{\beta_1 \beta_2 \cdots \beta_k} \tag{1.16}$$

holds in any coordinate system, although the components of the tensor appearing on both sides change. If a tensor equation is satisfied in a coordinate system, it is satisfied in any other coordinate system. Equivalently, if all the components of a tensor vanish in a coordinate system, then they vanish in all coordinate systems.

If a tensor equation contains a free index in its left hand side, then a free index with the same name and in the same position must appear on the right hand side. For example, $T_{\alpha\beta} A^\gamma B^\rho = X_{\alpha\beta}{}^{\gamma\rho}$ is a meaningful equation, while $A^\rho = B^\alpha$ and $A^\rho B_\mu = C^{\rho\mu}$ are meaningless.

The equations of physics are expressed as tensor equations to guarantee that they have the same form in all coordinate systems and they are not tied to any particular observer or coordinates, which would spoil their general character (Principle of General Covariance).

1.1.2.3   *Tensor symmetries*

The order in which indices appear in a tensor is important: in general, $P_{\alpha\beta} \neq P_{\beta\alpha}$ and $T^{\alpha\beta}{}_\gamma \neq T^{\beta\alpha}{}_\gamma \neq T_\gamma{}^{\alpha\beta}$. Certain tensors, however, exhibit symmetries when two of their indices are swapped.

A tensor of rank 2 (covariant or contravariant) $T_{\alpha\beta}$ or $T^{\alpha\beta}$ is symmetric if

$$T_{\beta\alpha} = T_{\alpha\beta} \quad \text{or} \quad T^{\beta\alpha} = T^{\alpha\beta} \tag{1.17}$$

for all values of the indices α and β. It is antisymmetric or skew–symmetric if

$$T_{\beta\alpha} = -T_{\alpha\beta} \quad \text{or} \quad T^{\beta\alpha} = -T^{\alpha\beta}. \tag{1.18}$$

In a manifold of dimension n, a rank 2 symmetric tensor has at most $\dfrac{n(n+1)}{2}$ independent components, while an antisymmetric rank 2 tensor has at most $\dfrac{n(n-1)}{2}$ independent components (all its diagonal components vanish).

A two–index contravariant (or covariant) tensor can always be decomposed into symmetric and antisymmetric parts as

$$T_{\alpha\beta} = \frac{T_{\alpha\beta} + T_{\beta\alpha}}{2} + \frac{T_{\alpha\beta} - T_{\beta\alpha}}{2} \equiv T_{(\alpha\beta)} + T_{[\alpha\beta]}, \tag{1.19}$$

where

$$T_{(\alpha\beta)} \equiv \frac{T_{\alpha\beta} + T_{\beta\alpha}}{2} \tag{1.20}$$

and

$$T_{[\alpha\beta]} \equiv \frac{T_{\alpha\beta} - T_{\beta\alpha}}{2} \tag{1.21}$$

are the symmetric part and the antisymmetric part of $T_{\alpha\beta}$, respectively. This decomposition is trivial and unique.

In general, the symmetric and antisymmetric parts of a tensor $T_{\mu_1\mu_2 \ldots \mu_k}$ are

$$T_{(\mu_1\mu_2 \ldots \mu_k)} \equiv \frac{1}{k!} \; (\text{sum over all permutations of } \mu_1 \ldots \mu_k) \tag{1.22}$$

$$T_{[\mu_1\mu_2 \ldots \mu_k]} \equiv \frac{1}{k!} \; (\text{sum over all permutations with alternating sign}) \tag{1.23}$$

(remember that the number of permutations of k objects is $k!$). For a three–index tensor $T_{\mu\nu\rho}$, we have

$$T_{(\alpha\beta\gamma)} = \frac{1}{3!} \left(T_{\alpha\beta\gamma} + T_{\alpha\gamma\beta} + T_{\gamma\alpha\beta} + T_{\gamma\beta\alpha} + T_{\beta\gamma\alpha} + T_{\beta\alpha\gamma} \right) \tag{1.24}$$

and

$$T_{[\alpha\beta\gamma]} = \frac{1}{3!} \left(T_{\alpha\beta\gamma} - T_{\alpha\gamma\beta} + T_{\gamma\alpha\beta} - T_{\gamma\beta\alpha} + T_{\beta\gamma\alpha} - T_{\beta\alpha\gamma} \right) . \tag{1.25}$$

A totally symmetric [antisymmetric] tensor is equal to its symmetric [antisymmetric] part. Tensors of rank higher than 2 can possess symmetries with respect to pairs of indices, for example $T_{ijk} = T_{jik}$ or $T_{ijk} = -T_{ikj}$. The symmetry or antisymmetry of a tensor is invariant under coordinate transformations.

The notation denoting symmetrization or antisymmetrization of indices is used also for indices belonging to different tensors, for example:

$$A_{(\mu} B_{\nu)} \equiv \frac{A_\mu B_\nu + A_\nu B_\mu}{2} ,$$

$$X^{[\alpha} Y^{\beta]} \equiv \frac{X^\alpha Y^\beta - X^\beta Y^\alpha}{2} .$$

1.1.2.4 *Tensor fields*

Tensor algebra is defined trivially: the sum of two tensors and the multiplication of a tensor by a scalar are defined component by component, defining tensor algebra. The trace of a mixed tensor $T^\alpha{}_\beta$ of type $\begin{pmatrix} 1 \\ 1 \end{pmatrix}$ is the scalar

$$T \equiv T^\alpha{}_\alpha \equiv \delta^\beta_\alpha T^\alpha{}_\beta \,. \tag{1.26}$$

Any antisymmetric two–index tensor is trace–free.

It is more interesting to use tensor fields instead of constant vectors or tensors. Tensor fields are defined on open regions of the spacetime manifold and their components have values depending on the position, $T^{\alpha_1 \,\cdots\, \alpha_\ell}{}_{\beta_1 \,\cdots\, \beta_m} = T^{\alpha_1 \,\cdots\, \alpha_\ell}{}_{\beta_1 \,\cdots\, \beta_m} (\vec{x})$. A continuous tensor field on a region $\mathcal{D}$ is such that all its components $T^{\alpha_1 \,\cdots\, \alpha_\ell}{}_{\beta_1 \,\cdots\, \beta_k} (x^\mu)$ are continuous functions of the coordinates x^μ at all points of $\mathcal{D}$. Similarly, one defines tensor fields that are differentiable to any order. One then introduces tensor calculus by taking derivatives and gradients of sufficiently regular tensor fields. A common notation is

$$\frac{\partial A_{\alpha\beta}}{\partial x^\gamma} \equiv A_{\alpha\beta,\gamma} \,, \tag{1.27}$$

$$\frac{\partial T_{\alpha\beta}{}^\gamma}{\partial x^\delta} \equiv T_{\alpha\beta}{}^\gamma{}_{,\delta} \,, \tag{1.28}$$

etc. However, the partial derivative of a tensor is not a tensor. For example, compute how the partial derivative $\partial/\partial x^\gamma$ of a tensor $T^{\alpha_1 \,\cdots\, \alpha_\ell}{}_{\beta_1 \,\cdots\, \beta_m}$ transforms:

$$\frac{\partial T^{\alpha'_1 \,\cdots\, \alpha'_\ell}{}_{\beta'_1 \,\cdots\, \beta'_m}}{\partial x^{\gamma'}} = \frac{\partial x^\delta}{\partial x^{\gamma'}}$$

$$\frac{\partial}{\partial x^\delta} \left[\frac{\partial x^{\alpha'_1}}{\partial x^{\mu_1}} \cdots \frac{\partial x^{\alpha'_\ell}}{\partial x^{\mu_\ell}} \frac{\partial x^{\nu_1}}{\partial x^{\beta'_1}} \cdots \frac{\partial x^{\nu_m}}{\partial x^{\beta'_m}} T^{\mu_1 \,\cdots\, \mu_\ell}{}_{\nu_1 \,\cdots\, \nu_m} \right]$$

$$= \frac{\partial x^\delta}{\partial x^{\gamma'}} \left[\frac{\partial}{\partial x^\delta} \left(\frac{\partial x^{\alpha_1'}}{\partial x^{\mu_1}} \cdots \frac{\partial x^{\alpha'_\ell}}{\partial x^{\mu_\ell}} \frac{\partial x^{\nu_1}}{\partial x^{\beta'_1}} \cdots \frac{\partial x^{\nu_m}}{\partial x^{\beta'_m}} \right) \right] T^{\mu_1 \,\cdots\, \mu_\ell}{}_{\nu_1 \,\cdots\, \nu_m}$$

$$+ \frac{\partial x^\delta}{\partial x^{\gamma'}} \frac{\partial x^{\alpha'_1}}{\partial x^{\mu_1}} \cdots \frac{\partial x^{\alpha'_\ell}}{\partial x^{\mu_\ell}} \frac{\partial x^{\nu_1}}{\partial x^{\beta_1'}} \cdots \frac{\partial x^{\nu_m}}{\partial x^{\beta'_m}} \frac{\partial T^{\mu_1 \,\cdots\, \mu_\ell}{}_{\nu_1 \,\cdots\, \nu_m}}{\partial x^\delta} \,, \tag{1.29}$$

which deviates from the tensor transformation property because of the second last line, except for linear transformations for which $\dfrac{\partial^2 x^{\alpha'}}{\partial x^\beta \partial x^\delta}$ and $\dfrac{\partial^2 x^\alpha}{\partial x^{\beta'} \partial x^{\delta'}}$ vanish. The same is true for higher order partial derivatives of a tensor. According to the Principle of General Covariance, we must define a new notion of derivative ("covariant derivative") ∇_μ that, when applied to tensor fields, produces tensor fields.

1.1.3   *Metric tensor*

The metric tensor is a tensor field that provides the notion of distance between points (in the Lorentzian sense), of scalar product between vectors, and of length of a vector on the spacetime manifold.

A metric is a tensor field $g_{\alpha\beta}$ of type $\begin{pmatrix} 0 \\ 2 \end{pmatrix}$ which is symmetric and non–degenerate, *i.e.*,

$$g_{\alpha\beta} X^\alpha Y^\beta = 0 \qquad \forall\, X^\alpha \Rightarrow Y^\alpha = 0 \,. \tag{1.30}$$

The infinitesimal distance squared between two points of coordinates x^μ and $x^\mu + dx^\mu$, or line element (squared), is

$$ds^2 = g_{\mu\nu}\, dx^\mu dx^\nu \,. \tag{1.31}$$

For example, the metric of the Minkowski spacetime of special relativity is given by [136, 250, 342, 435]

$$ds^2 = -dt^2 + dx^2 + dy^2 + dz^2 \tag{1.32}$$

in Cartesian coordinates $\left(t, x, y, z \right)$ and the Schwarzschild metric describing gravity in the empty space outside a spherical black hole in general relativity is given by [82, 120, 202, 435]

$$ds^2 = -\left(1 - \frac{2m}{r}\right) dt^2 + \frac{dr^2}{1 - 2m/r} + r^2 \left(d\vartheta^2 + \sin^2 \vartheta \, d\varphi^2 \right) \tag{1.33}$$

in polar coordinates $\left(t, r, \vartheta, \varphi \right)$, where the constant m is the black hole mass.

The scalar product between two vectors A^μ and B^μ is

$$g_{\mu\nu} A^\mu B^\nu \,. \tag{1.34}$$

Two vectors are said to be orthogonal if their scalar product vanishes.

As a special case, if $B^\mu = A^\mu$, one has the notion of length squared of a vector

$$g_{\mu\nu} A^\mu A^\nu \,. \tag{1.35}$$

A vector A^μ is timelike if $g_{\mu\nu}A^\mu A^\nu < 0$, null or lightlike if $g_{\mu\nu}A^\mu A^\nu = 0$, or spacelike if $g_{\mu\nu}A^\mu A^\nu > 0$.

For example, in $\mathbb{R}^n$ with Cartesian coordinates $\{x^1, \dots, x^n\}$, the Euclidean metric is

$$e_{\mu\nu} = \delta_{\mu\nu} \qquad (1.36)$$

and the infinitesimal distance squared between two points x^μ and $x^\mu + dx^\mu$ is

$$d\ell_{(n)}^2 = \delta_{\mu\nu}\, dx^\mu dx^\nu = \left(dx^1\right)^2 + \left(dx^2\right)^2 + \dots + \left(dx^n\right)^2 , \qquad (1.37)$$

which expresses the Pythagorean theorem in n dimensions. The Euclidean metric is positive–definite since $d\ell_{(n)}^2 \geq 0$ and $d\ell_{(n)} = 0$ if and only if $\left(dx^1, \dots, dx^n\right) = (0, \dots, 0)$. The scalar product between two vectors defined by $e_{\mu\nu}$ is the usual dot product. In Cartesian coordinates, we have

$$e_{\mu\nu}A^\mu B^\nu = \delta_{\mu\nu}A^\mu B^\nu = A^1 B^1 + A^2 B^2 + \dots + A^n B^n = \vec{A} \cdot \vec{B} \qquad (1.38)$$

and the length squared of a vector $\vec{A} = \left(A^1, \dots, A^n\right)$ is

$$\|A\|^2 = e_{\mu\nu}A^\mu A^\nu = \sum_{\mu=1}^{n} (A^\mu)^2 = \vec{A} \cdot \vec{A}. \qquad (1.39)$$

If the metric tensor is represented by a matrix with components $g_{\mu\nu}$ in some coordinates, the inverse metric always exists and is given by the inverse matrix with components $g^{\mu\nu}$. Since $g_{\mu\nu}$ is symmetric, so is $g^{\mu\nu}$. The tensor represented by the matrix $(g^{\mu\nu})$ is called the inverse metric and satisfies

$$g_{\mu\nu}\, g^{\nu\alpha} = \delta_\mu^\alpha . \qquad (1.40)$$

The mixed components of the metric tensor coincide with the Kronecker delta, $g^\mu{}_\nu = \delta_\nu^\mu$.

The metric tensor allows one to raise and lower tensor indices. If A^μ is a contravariant vector, we define

$$A_\mu \equiv g_{\mu\nu}A^\nu \qquad (\text{``lowering'' of } \mu) \qquad (1.41)$$

and, if B_μ is a covariant vector, we define

$$B^\mu \equiv g^{\mu\nu} B_\nu \qquad (\text{``raising'' of } \mu). \qquad (1.42)$$

The raising of a lowered index and the lowering of a raised index give back the original object:

$$A^\mu = g^{\mu\nu} A_\nu = g^{\mu\nu} \left(g_{\nu\alpha}A^\alpha\right) = \delta_\alpha^\mu A^\alpha = A^\mu ,$$
$$B_\mu = g_{\mu\nu}B^\nu = g_{\mu\nu} \left(g^{\nu\beta} B_\beta\right) = \delta_\mu^\beta B_\beta = B_\mu . \qquad (1.43)$$

The operations of raising or lowering apply to any tensor index, for example $T^{\alpha}{}_{\beta\gamma} = g^{\alpha\mu} T_{\mu\beta\gamma}$ and $Q^{\mu\nu} = g^{\mu\alpha} g^{\nu\beta} Q_{\alpha\beta}$. A^{μ} and A_{μ} are just different descriptions of the same object: it is always possible to obtain one from the other by lowering or raising indices with the metric or the inverse metric. The scalar product of two vectors A^{μ} and B^{μ} can now be written as

$$g_{\mu\nu} A^{\mu} B^{\nu} = g^{\mu\nu} A_{\mu} B_{\nu} = A_{\mu} B^{\mu} = A^{\mu} B_{\mu} \,. \qquad (1.44)$$

Raising or lowering the indices of a tensor preserves its symmetries.

The metric determinant g is simply the determinant of the matrix that represents it in a coordinate system, $g \equiv \mathrm{Det}\,(g_{\mu\nu})$. Since the metric tensor is non–degenerate, the associated matrix is non–singular and $g \neq 0$.

In the following, a spacetime $(\mathcal{M}, g_{\mu\nu})$ is described by a four–dimensional manifold $\mathcal{M}$ on which a metric tensor field $g_{\mu\nu}$ with Lorentzian signature $-+++$ is defined. We follow the notations and conventions of Ref. [435]. We use units in which the speed of light c and Newton's constant G are unity, but these constants are occasionally restored for clarity.

1.1.4   *Levi–Civita symbol and tensor densities*

The Levi–Civita, or alternating, symbol in three dimensions is defined as

$$\varepsilon_{\mu\nu\rho} \equiv \begin{cases} +1 & \text{if } \mu\nu\rho \text{ is an even permutation of } 123, \\ -1 & \text{if } \mu\nu\rho \text{ is an odd permutation of } 123, \\ 0 & \text{otherwise,} \end{cases} \qquad (1.45)$$

where a permutation of 123 is an ordering of the set $\left\{1, 2, 3\right\}$ obtained by starting with the natural order 123 and exchanging only two digits at a time, once or several times. An even (odd) permutation is one that requires an even (odd) number of such exchanges. There are $n!$ permutations of n numbers, therefore the non–zero components of the Levi–Civita symbol $\varepsilon_{\mu\nu\rho}$ are

$$\varepsilon_{123} = +1\,, \ \varepsilon_{132} = -1\,, \ \varepsilon_{312} = +1\,,$$
$$\varepsilon_{321} = -1\,, \ \varepsilon_{231} = +1\,, \ \varepsilon_{213} = -1\,.$$

In four dimensions $n = 4$, the Levi–Civita symbol is defined as

$$\varepsilon_{\mu\nu\rho\sigma} \equiv \begin{cases} +1 & \text{if } \mu\nu\rho\sigma \text{ is an even permutation of } 0123, \\ -1 & \text{if } \mu\nu\rho\sigma \text{ is an odd permutation of } 0123, \\ 0 & \text{otherwise,} \end{cases} \qquad (1.46)$$

where a permutation of 0123 is an ordering of the set $\{0, 1, 2, 3\}$ obtained by starting with the natural order 0123 and exchanging only two digits at a time. Now there are $4! = 24$ permutations of four indices, giving the 24 components

$$
\begin{aligned}
&\varepsilon_{0123} = +1\,,\; \varepsilon_{0132} = -1\,,\; \varepsilon_{0213} = -1\,, \\
&\varepsilon_{0231} = +1\,,\; \varepsilon_{0312} = +1\,,\; \varepsilon_{0321} = -1\,, \\
&\varepsilon_{1023} = -1\,,\; \varepsilon_{1032} = +1\,,\; \varepsilon_{1203} = +1\,, \\
&\varepsilon_{1230} = -1\,,\; \varepsilon_{1302} = -1\,,\; \varepsilon_{1320} = +1\,, \\
&\varepsilon_{2013} = +1\,,\; \varepsilon_{2031} = -1\,,\; \varepsilon_{2103} = -1\,, \\
&\varepsilon_{2130} = +1\,,\; \varepsilon_{2301} = +1\,,\; \varepsilon_{2310} = -1\,, \\
&\varepsilon_{3012} = -1\,,\; \varepsilon_{3021} = +1\,,\; \varepsilon_{3102} = +1\,, \\
&\varepsilon_{3120} = -1\,,\; \varepsilon_{3201} = -1\,,\; \varepsilon_{3210} = +1\,.
\end{aligned}
$$

The Levi–Civita symbol transforms as a tensor only in $\mathbb{R}^n$ in Cartesian coordinates, but not in other coordinate systems. It is not a tensor because it has the same components in any coordinate system: it is a tensor density.

A tensor density in four dimensions $\tau_{\mu_1 \mu_2 \mu_3 \mu_4}$ is a quantity that transforms according to

$$
\tau_{\mu_1' \mu_2' \mu_3' \mu_4'} = \left[\mathrm{Det}\left(\frac{\partial x^{\mu'}}{\partial x^{\mu}} \right) \right]^w \tau_{\mu_1 \mu_2 \mu_3 \mu_4} \frac{\partial x^{\mu_1}}{\partial x^{\mu_1'}} \frac{\partial x^{\mu_2}}{\partial x^{\mu_2'}} \frac{\partial x^{\mu_3}}{\partial x^{\mu_3'}} \frac{\partial x^{\mu_4}}{\partial x^{\mu_4'}} \tag{1.47}
$$

under the coordinate change $x^{\mu} \to x^{\mu'}(x^{\alpha})$, where w is the weight of the density.

One can always construct a tensor from a tensor density $\tau_{\mu\nu\rho\sigma}$ of weight w by multiplying it by the power of the absolute value of the metric determinant $|g|^{w/2}$:

$$
t_{\mu\nu\rho\sigma} \equiv \left(\sqrt{|g|} \right)^w \tau_{\mu\nu\rho\sigma} \tag{1.48}
$$

is a true tensor.

It can be shown (*e.g.*, [120]) that the Levi–Civita symbol is a tensor density of weight $+1$, while the metric determinant g is a (scalar) density of weight -2.

Although the Levi–Civita symbol $\varepsilon_{\mu\nu\rho\sigma}$ transforms as a tensor density, the quantity

$$
\tilde{\varepsilon}_{\mu\nu\rho\sigma} \equiv \sqrt{|g|}\, \varepsilon_{\mu\nu\rho\sigma}\,, \tag{1.49}
$$

transforms as a true tensor [82, 120, 435].

In four spacetime dimensions, the Levi–Civita symbol has the following properties.

By raising indices with the inverse metric, we define $\varepsilon^{\mu\nu\rho\sigma}$ and

$$\tilde{\varepsilon}^{\mu\nu\rho\sigma} = \frac{1}{\sqrt{|g|}}\,\varepsilon^{\mu\nu\rho\sigma}\,; \tag{1.50}$$

then contraction yields

$$\tilde{\varepsilon}^{\mu\nu\rho\sigma}\,\tilde{\varepsilon}_{\mu\nu\rho\sigma} = -4! = -24\,, \tag{1.51}$$

$$\tilde{\varepsilon}^{\mu\nu\rho\alpha}\,\tilde{\varepsilon}_{\mu\nu\rho\beta} = -3!\,1!\,\delta^{\alpha}_{\beta} = -6\,\delta^{\alpha}_{\beta}\,, \tag{1.52}$$

$$\tilde{\varepsilon}^{\mu\nu\alpha\beta}\,\tilde{\varepsilon}_{\mu\nu\gamma\delta} = -2!\,2!\,\delta^{[\alpha}_{\gamma}\,\delta^{\beta]}_{\delta} = -2\left(\delta^{\alpha}_{\gamma}\,\delta^{\beta}_{\delta} - \delta^{\beta}_{\gamma}\,\delta^{\alpha}_{\delta}\right)\,, \tag{1.53}$$

$$\tilde{\varepsilon}^{\mu\nu\alpha\beta}\,\tilde{\varepsilon}_{\mu\gamma\delta\varphi} = -3!\,1!\,\delta^{[\nu}_{\gamma}\,\delta^{\alpha}_{\delta}\,\delta^{\beta]}_{\varphi}\,. \tag{1.54}$$

Finally, if $A_{\mu\nu}$ is an antisymmetric two–index tensor, its "dual" is the tensor density

$$^{*}A_{\mu\nu} \equiv \frac{1}{2}\,\varepsilon_{\mu\nu}{}^{\rho\sigma}\,A_{\rho\sigma}\,. \tag{1.55}$$

As is well known, the volume element $d^n x$ in n dimensions does not transform as a tensor under coordinate transformations, but as a tensor density because it involves the Jacobian of the transformation:

$$d^n x' = \mathrm{Det}\left(\frac{\partial x^{\alpha'}}{\partial x^{\beta}}\right) d^n x\,. \tag{1.56}$$

Since g is a tensor density of weight -2, we can use $\sqrt{|g|}$ to construct a covariant volume element. The quantity $\sqrt{|g|}\,d^n x$ is invariant,

$$\sqrt{|g'|}\,d^n x' = \sqrt{|g|}\,d^n x\,; \tag{1.57}$$

this is the correct volume element used in relativity. A covariant expression of the integral of a function $f(x^\mu)$ over a region A of the spacetime manifold is

$$\int_A d^n x\,\sqrt{|g|}\,f \tag{1.58}$$

and not $\int_A d^n x\,f$. By the same token, the volume element $\sqrt{|g|}\,d^n x$ enters the integral of tensorial quantities,

$$\int_\Omega d^n x\,\sqrt{|g|}\,T^{\mu_1\mu_2\,\cdots\,\mu_k}{}_{\nu_1\nu_2\,\ldots\,\nu_\ell}\,. \tag{1.59}$$

1.1.5 *Covariant derivative*

To begin doing physics on curved manifolds, one needs to compute rates of change of quantities in space and time. The notion of partial derivative is not sufficient to deal with the calculations occurring in differential geometry and general relativity because, when applied to tensors, it depends on the particular coordinates adopted. It needs to be replaced by the notion of covariant derivative, which is introduced by the following axioms.[2]

Let $\tau(k,\ell)$ denote the space of smooth tensor fields of type $\begin{pmatrix} k \\ \ell \end{pmatrix}$ on the spacetime manifold M and let $\mathcal{F} = \tau(0,0)$ be the space of smooth functions defined on an open region $O \subseteq M$.

A *covariant derivative operator* is an operator

$$\nabla_\mu : \quad \tau(k,\ell) \longrightarrow \tau(k,\ell+1) \tag{1.60}$$

that takes each differentiable tensor field of type $\begin{pmatrix} k \\ \ell \end{pmatrix}$ into a tensor field of type $\begin{pmatrix} k \\ \ell+1 \end{pmatrix}$ and satisfies the following properties:

(1) *Linearity:*
$$\forall A^{\alpha_1 \cdots \alpha_k}{}_{\beta_1 \ldots \beta_\ell}, B^{\alpha_1 \cdots \alpha_k}{}_{\beta_1 \ldots \beta_\ell} \in \tau(k,\ell), \quad \forall a,b \text{ real,}$$

$$\nabla_\mu(aA^{\alpha_1 \cdots \alpha_k}{}_{\beta_1 \ldots \beta_\ell} + bB^{\alpha_1 \cdots \alpha_k}{}_{\beta_1 \ldots \beta_\ell}) = a\nabla_\mu A^{\alpha_1 \cdots \alpha_k}{}_{\beta_1 \ldots \beta_\ell}$$

$$+ b\nabla_\mu B^{\alpha_1 \cdots \alpha_k}{}_{\beta_1 \ldots \beta_\ell}$$

(2) *Leibnitz property:*
$$\forall A^{\alpha_1 \cdots \alpha_k}{}_{\beta_1 \ldots \beta_\ell} \in \tau(k,\ell), \quad \forall B^{\alpha_1 \cdots \alpha_{k'}}{}_{\beta_1 \ldots \beta_{\ell'}} \in \tau(k',\ell'),$$
$$\nabla_\mu\left(A^{\alpha_1 \cdots \alpha_k}{}_{\beta_1 \ldots \beta_\ell} B^{\alpha_1 \cdots \alpha_{k'}}{}_{\beta_1 \ldots \beta_{\ell'}}\right)$$

$$= \left(\nabla_\mu A^{\alpha_1 \cdots \alpha_k}{}_{\beta_1 \ldots \beta_\ell}\right) B^{\alpha_1 \cdots \alpha_{k'}}{}_{\beta_1 \ldots \beta_{\ell'}}$$

$$+ A^{\alpha_1 \cdots \alpha_k}{}_{\beta_1 \ldots \beta_\ell} \nabla_\mu B^{\alpha_1 \cdots \alpha_{k'}}{}_{\beta_1 \ldots \beta_{\ell'}}$$

(3) *Commutativity with contraction:*
$$\forall A^{\alpha_1 \cdots \alpha_k}{}_{\beta_1 \ldots \beta_\ell} \in \tau(k,\ell),$$
$$\nabla_\nu\left(A^{\alpha_1 \cdots \mu \cdots \alpha_k}{}_{\beta_1 \ldots \mu \ldots \beta_\ell}\right) = \nabla_\nu A^{\alpha_1 \cdots \mu \cdots \alpha_k}{}_{\beta_1 \ldots \mu \ldots \beta_\ell}$$

[2]This definition applies to the flat Minkowski spacetime of special relativity in curvilinear coordinates as well.

(4) *Consistency with the notion of directional derivative on a scalar:*
$\forall f \in \mathcal{F},$ for any vector field X^μ,

$$X(f) \equiv X^\mu \partial_\mu f = X^\alpha \nabla_\alpha f$$

(5) *Torsion–free property:*
$\forall f \in \mathcal{F},$

$$\nabla_\alpha \nabla_\beta f = \nabla_\beta \nabla_\alpha f$$

(second covariant derivatives of a scalar function commute).

Infinitely many covariant derivative operators can be defined on a manifold. However, if a metric tensor $g_{\alpha\beta}$ is defined on M, it determines uniquely a preferred covariant derivative operator ("metric connection"): the one that satisfies the property

$$\nabla_\mu g_{\alpha\beta} = 0. \tag{1.61}$$

We will always have a metric tensor $g_{\mu\nu}$ defined on the spacetime manifold and the covariant derivative used will always be the one defined by the metric connection induced by $g_{\mu\nu}$.

1.1.6   *Connection, geodesics, and curvature*

In practice, the metric connection is expressed by the Christoffel symbols (or connection coefficients)

$$\Gamma^\mu_{\alpha\beta} = \frac{1}{2} g^{\mu\sigma} \left(g_{\sigma\alpha,\beta} + g_{\sigma\beta,\alpha} - g_{\alpha\beta,\sigma} \right) \tag{1.62}$$

which are symmetric in their lower indices,

$$\Gamma^\mu_{\alpha\beta} = \Gamma^\mu_{\beta\alpha} \tag{1.63}$$

by definition.

The $\Gamma^\mu_{\alpha\beta}$ are not tensors since they do not transform as such under coordinate transformations. Their transformation property under a coordinate change $x^\mu \longrightarrow x^{\mu'}$ is instead

$$\Gamma^{\alpha'}_{\mu'\nu'} = \frac{\partial x^{\alpha'}}{\partial x^\delta} \frac{\partial x^\rho}{\partial x^{\mu'}} \frac{\partial x^\sigma}{\partial x^{\nu'}} \Gamma^\delta_{\rho\sigma} + \frac{\partial x^{\alpha'}}{\partial x^\delta} \frac{\partial^2 x^\delta}{\partial x^{\mu'} \partial x^{\nu'}}. \tag{1.64}$$

However, the Γ–symbols are used to calculate true tensors. The covariant derivative of a contravariant vector field A^μ is

$$\nabla_\mu A^\alpha = \partial_\mu A^\alpha + \Gamma^\alpha_{\mu\nu} A^\nu, \tag{1.65}$$

i.e., the ordinary derivative plus an additional term constructed with the connection coefficients and the vector field itself. In this additional term the contravariant index α of the four–vector is taken by the Γ–symbol. The positive sign with which the second term on the right hand side enters the expression of $\nabla_\mu A^\alpha$ is associated with the contravariant nature of the index α in A^α. By contrast, the covariant derivative of a covariant vector field B_μ is

$$\nabla_\mu B_\alpha = \partial_\mu B_\alpha - \Gamma^\nu_{\mu\alpha} B_\nu \qquad (1.66)$$

with a negative sign of the second term on the right hand side. Using Eq. (1.64), it is straightforward to verify that $\nabla_\mu A^\alpha$ transforms as a tensor of type $\begin{pmatrix} 1 \\ 1 \end{pmatrix}$ and $\nabla_\mu B_\alpha$ transforms as a tensor of type $\begin{pmatrix} 0 \\ 2 \end{pmatrix}$.

The rules for computing covariant derivatives extend to all covariant and contravariant indices of a tensor field $T^{\alpha_1 \ldots \alpha_\ell}{}_{\beta_1 \ldots \beta_m}$:

$$\nabla_\mu T^{\alpha_1 \ldots \alpha_\ell}{}_{\beta_1 \ldots \beta_m} = \partial_\mu T^{\alpha_1 \ldots \alpha_\ell}{}_{\beta_1 \ldots \beta_m}$$

$$+\Gamma^{\alpha_1}_{\mu\sigma} T^{\sigma\alpha_2 \ldots \alpha_\ell}{}_{\beta_1 \ldots \beta_m}$$

$$+\Gamma^{\alpha_2}_{\mu\sigma} T^{\alpha_1\sigma\alpha_2 \ldots \alpha_\ell}{}_{\beta_1 \ldots \beta_m} + \ldots \left.\begin{array}{l} \\ \\ \\ \end{array}\right\} \begin{array}{l} \ell \text{ terms for the} \\ \ell \text{ contravariant} \\ \text{indices} \end{array}$$

$$+\Gamma^{\alpha_\ell}_{\mu\sigma} T^{\alpha_1\alpha_2\ldots\sigma}{}_{\beta_1 \ldots \beta_m}$$

$$\qquad (1.67)$$

$$-\Gamma^{\sigma}_{\mu\beta_1} T^{\alpha_1 \ldots \alpha_\ell}{}_{\sigma\beta_2 \ldots \beta_m}$$

$$-\Gamma^{\sigma}_{\mu\beta_2} T^{\alpha_1 \ldots \alpha_\ell}{}_{\beta_1\sigma\beta_2 \ldots \beta_m} - \ldots \left.\begin{array}{l} \\ \\ \\ \end{array}\right\} \begin{array}{l} m \text{ terms for the} \\ m \text{ covariant} \\ \text{indices} \end{array}$$

$$-\Gamma^{\sigma}_{\mu\beta_m} T^{\alpha_1 \ldots \alpha_\ell}{}_{\beta_1\beta_2 \ldots \sigma} .$$

Again, using Eqs. (1.64) and (1.67), it is straightforward to check that the covariant derivatives $\nabla_\mu T^{\alpha_1 \ldots \alpha_\ell}{}_{\beta_1 \ldots \beta_m}$ transform as true tensors under coordinate transformations.

The notion of covariant derivative (or connection) leads to that of geodesic curve. Let $x^\mu(\lambda)$ be a spacetime curve parametrized by a parameter λ and with four–tangent $u^\mu \equiv dx^\mu/d\lambda$. We are interested in timelike and null curves, which are the four–trajectories of physical particles. For timelike curves we adopt the parameter $\lambda = \tau$, where τ is the proper time along the curve and the tangent to the curve is the four–velocity

$$u^\mu = \frac{dx^\mu}{d\tau}, \qquad u_\mu u^\mu = -1. \qquad (1.68)$$

The derivative of a tensor field $T^{\mu_1 \cdots \mu_k}{}_{\nu_1 \ldots \nu_\ell}$ along the curve is defined as

$$\frac{DT^{\mu_1 \cdots \mu_k}{}_{\nu_1 \ldots \nu_\ell}}{D\lambda} \equiv u^\alpha \nabla_\alpha T^{\mu_1 \cdots \mu_k}{}_{\nu_1 \ldots \nu_\ell} , \tag{1.69}$$

i.e., as the projection of the covariant derivative of the tensor along the direction of the curve (the direction of its tangent).

A geodesic curve is a curve of zero four–dimensional acceleration, or

$$u^\beta \nabla_\beta u^\mu = 0 . \tag{1.70}$$

This definition requires that the variation of the tangent u^μ in the direction of the curve vanishes. Since

$$\frac{Du^\mu}{D\lambda} \equiv u^\beta \nabla_\beta u^\mu = u^\beta \left(\partial_\beta u^\mu + \Gamma^\mu_{\beta\alpha} u^\alpha \right)$$

$$= \frac{dx^\beta}{d\lambda} \frac{\partial u^\mu}{\partial x^\beta} + \Gamma^\mu_{\beta\alpha} u^\alpha u^\beta$$

$$= \frac{du^\mu}{d\lambda} + \Gamma^\mu_{\beta\alpha} u^\alpha u^\beta ,$$

a geodesic curve obeys the geodesic equation

$$\frac{Du^\mu}{D\lambda} = u^\beta \nabla_\beta u^\mu = 0 \tag{1.71}$$

equivalent to

$$\frac{d^2 x^\mu}{d\lambda^2} + \Gamma^\mu_{\alpha\beta} \frac{dx^\alpha}{d\lambda} \frac{dx^\beta}{d\lambda} = 0 . \tag{1.72}$$

In $\mathbb{R}^n$ and in Minkowski spacetime, in Cartesian coordinates the connection coefficients vanish identically and the geodesic equation reduces to

$$\frac{du^\mu}{d\lambda} = \frac{d^2 x^\mu}{d\lambda^2} = 0 , \tag{1.73}$$

which has the solution $x^\mu(\lambda) = A^\mu \lambda + B^\mu$, where A^μ and B^μ are constants: the geodesics of Minkowski spacetime are straight lines.

A property of geodesics is that they extremize the length between two fixed points [42, 183, 284, 438]

$$\ell_{12} = \int_1^2 \sqrt{-g_{\alpha\beta} u^\alpha u^\beta} \, d\lambda , \tag{1.74}$$

where we consider timelike curves. It is well known that the minimum length between two points in $\mathbb{R}^n$ with Euclidean metric is obtained by following the straight line joining them. In curved space, geodesics still extremize the length between two fixed points defined using the metric $g_{\mu\nu}$ instead of the Euclidean metric.

The form of the geodesic equation can be more general than (1.70). A geodesic can be defined more generally as a curve such that its tangent is transported parallel to itself when moving along the geodesic, that is

$$\frac{Du^\mu}{D\lambda} \equiv u^\nu \nabla_\nu u^\mu = \alpha \, u^\mu \qquad (1.75)$$

where α is a constant, or

$$\frac{d^2 x^\mu}{d\lambda^2} + \Gamma^\mu_{\rho\sigma} \frac{dx^\rho}{d\lambda} \frac{dx^\sigma}{d\lambda} = \alpha \frac{dx^\mu}{d\lambda} \, . \qquad (1.76)$$

This equation expresses the fact that the change of the tangent vector u^μ in the direction of the curve is parallel to u^μ itself, *i.e.*, this vector is transported parallel to itself when moving along the curve.

As for all curves, many parametrizations can be chosen for the same geodesic. A particular class of parameters (affine parameters) is such that Eq. (1.75) reduces to the simpler form (1.70), which is said to be the affinely parametrized form. Non–affinely parametrized geodesics are still curves of extremal length, only their parameters are non–affine.

The curvature is defined by using the fact that, in a curved space, transporting a vector parallel to itself along an infinitesimal loop depends on the direction of the travel. The failure of obtaining the same result when transporting this vector from an initial to a final point along the two possible paths in the same loop measures the local curvature. The latter is expressed by the Riemann or curvature tensor $\mathcal{R}_{\alpha\beta\gamma}{}^\delta$ defined by [82, 202, 250, 435]

$$[\nabla_\alpha, \nabla_\beta] \, \omega_\gamma = \mathcal{R}_{\alpha\beta\gamma}{}^\delta \omega_\delta \, . \qquad (1.77)$$

We also have

$$[\nabla_\alpha, \nabla_\beta] \, V^\gamma = -\mathcal{R}_{\alpha\beta\delta}{}^\gamma V^\delta \, . \qquad (1.78)$$

The Riemann tensor satisfies

$$\mathcal{R}_{\alpha\beta\gamma}{}^{\delta} = -\mathcal{R}_{\beta\alpha\gamma}{}^{\delta}, \tag{1.79}$$

$$\mathcal{R}_{\alpha\beta\gamma\delta} = -\mathcal{R}_{\alpha\beta\delta\gamma}, \tag{1.80}$$

$$\mathcal{R}_{\alpha\beta\gamma\delta} = \mathcal{R}_{\gamma\delta\alpha\beta}, \tag{1.81}$$

$$\mathcal{R}_{[\alpha\beta\gamma]}{}^{\delta} = 0, \tag{1.82}$$

$$\nabla_{[\alpha}\mathcal{R}_{\beta\gamma]\delta}{}^{\mu} = 0 \qquad \text{(Bianchi identity)}. \tag{1.83}$$

Once a coordinate system is fixed, the Riemann tensor is calculated using the Christoffel symbols and their derivatives as [82, 435]

$$\mathcal{R}_{\alpha\beta\gamma}{}^{\delta} = \Gamma^{\delta}_{\alpha\gamma,\beta} - \Gamma^{\delta}_{\beta\gamma,\alpha} + \Gamma^{\sigma}_{\alpha\gamma}\Gamma^{\delta}_{\sigma\beta} - \Gamma^{\sigma}_{\beta\gamma}\Gamma^{\delta}_{\sigma\alpha}. \tag{1.84}$$

The Ricci tensor is the contraction

$$\mathcal{R}_{\alpha\gamma} = \mathcal{R}_{\alpha\beta\gamma}{}^{\beta}, \tag{1.85}$$

the Ricci scalar is

$$\mathcal{R} \equiv \mathcal{R}^{\sigma}{}_{\sigma} = g^{\alpha\beta}\mathcal{R}_{\alpha\beta}, \tag{1.86}$$

and the Einstein tensor is

$$G_{\mu\nu} \equiv R_{\mu\nu} - \frac{1}{2}g_{\mu\nu}R. \tag{1.87}$$

As a consequence of Eq. (1.83), the Einstein tensor satisfies the contracted Bianchi identity

$$\nabla^{\nu}G_{\mu\nu} = 0. \tag{1.88}$$

1.1.7 *Energy–momentum tensors*

In relativity, a mass–energy distribution (which can be a material, a fluid, or a field) is described by a symmetric two–index tensor $T_{\mu\nu}$ ("stress–energy" or "energy–momentum" tensor) that contains as its various components the energy density, energy flux density, isotropic pressure, and stresses of this matter. In general, the stress–energy tensor cannot be completely specified *a priori* because it contains the metric $g_{\mu\nu}$, which is not known until the field equations of the theory of gravity (usually Einstein theory) are solved.

The stress–energy tensor $T_{\mu\nu}$ satisfies the covariant conservation equation

$$\nabla^\nu T_{\mu\nu} = 0\,, \tag{1.89}$$

which contains the equations of motion of the matter or field described by $T_{\mu\nu}$. This includes the relativistic generalization of the Navier–Stokes equations for a fluid, the Klein–Gordon equation for a scalar field, the Maxwell equations for the electromagnetic field, *etc.*

1.1.7.1 *Perfect fluids*

Perfect fluids have no dissipation or viscosity, no heat conduction, and are characterized by a timelike four–velocity field u^μ satisfying the usual normalization

$$u_\mu u^\mu = -1\,, \tag{1.90}$$

an energy density ρ, and an isotropic pressure P. The corresponding stress–energy tensor has the form

$$T_{\mu\nu} = (P + \rho)\,u_\mu u_\nu + P g_{\mu\nu}\,. \tag{1.91}$$

The pressure and energy density are related by an equation of state, *i.e.*, a functional relation $f(\rho, P, T) = 0$ between energy density, pressure, and temperature T. In cosmology we are interested in barotropic equations of state of the form $P = P(\rho)$ and, most often, in linear and constant equations of state

$$P = w\rho \tag{1.92}$$

with constant equation of state parameter w. Sometimes the adiabatic index γ defined by $w \equiv \gamma - 1$ is used. Time–dependent equations of state with $w = w(t)$ are also used.

The equation of state (1.92) covers several fluids of importance to cosmology:

- $P = \rho/3$, corresponding to $w = 1/3$ and $\gamma = 4/3$, describes blackbody radiation;
- $P = 0$, corresponding to $w = 0$ and $\gamma = 1$, describes non–relativistic (cold) matter, called "dust";
- $P = -\rho$, or $w = -1$ and $\gamma = 0$, describes the quantum vacuum or the cosmological constant;
- $P = \rho$, corresponding to $w = 1$ and $\gamma = 2$, is associated with a "stiff fluid", which is realized also by a free scalar field;
- $-\rho \leq P < -\rho/3$, or $-1 \leq w < -1/3$, is referred to as "dark energy";
- $P < -\rho$, or $w < -1$ and $\gamma < 0$, is called "phantom fluid".

1.1.7.2   Scalar field

Scalar fields [172] are ubiquitous in general relativity, cosmology, and particle physics. Scalar fields are usually assumed to drive inflation in the early
universe and, in many models of dark energy, they drive the present–day
acceleration of the cosmic expansion (scalar field dark energy is referred to
as "quintessence").

In field theory parlance, scalar fields correspond to spin zero particles
and occur in many theories. In the Standard Model of particle physics,
the Higgs field that breaks electroweak symmetry and gives particles their
masses is a fundamental scalar field, detected at the Large Hadron Collider
at *CERN*. In speculative theories beyond the Standard Model, scalar fields
break symmetries of supersymmetric or grand–unified theories.

The stress energy tensor of a classical scalar field ϕ is

$$T_{\mu\nu}^{(\phi)} = \nabla_\mu\phi\nabla_\nu\phi - \frac{1}{2}g_{\mu\nu}\nabla^\alpha\phi\nabla_\alpha\phi - Vg_{\mu\nu} \tag{1.93}$$

where $V(\phi)$ is the potential energy density ("the potential"). The covariant
conservation equation $\nabla^\nu T_{\mu\nu}^{(\phi)} = 0$ reproduces the Klein–Gordon equation

$$\Box\phi - \frac{dV}{d\phi} = 0 \,. \tag{1.94}$$

If a scalar field ϕ has timelike four–gradient, $\nabla^\mu\phi\nabla_\mu\phi < 0$, its stress
energy tensor is equivalent to that of an effective perfect fluid with four–
velocity

$$u^\mu = \frac{\nabla^\mu\phi}{\sqrt{|\nabla^\alpha\phi\nabla_\alpha\phi|}} \tag{1.95}$$

and energy density and pressure

$$\rho_{(\phi)} = -\frac{1}{2}\nabla^\mu\phi\nabla_\mu\phi + V(\phi) \,, \tag{1.96}$$

$$P_{(\phi)} = -\frac{1}{2}\nabla^\mu\phi\nabla_\mu\phi - V(\phi) \,. \tag{1.97}$$

Clearly, a free (*i.e.*, $V(\phi) = 0$) scalar field has the equation of state
$P_{(\phi)} = \rho_{(\phi)}$ of a stiff fluid.

1.1.7.3   Energy conditions

Here we summarize the point–wise energy conditions of relativistic gravity;
there are slight differences in the definitions of some of them in the literature
(*e.g.*, [202, 435]). The energy conditions satisfied by a specific form of matter

are formulated in terms of the stress–energy tensor $T_{\mu\nu}$ that describes it. Since we will mostly use perfect fluids in this book, we specify the energy conditions to the form (1.91) of the stress–energy tensor.

The *weak energy condition* (WEC) consists of

$$T_{\alpha\beta}\, t^\alpha\, t^\beta \geq 0 \qquad \text{for all timelike vectors } t^\mu. \tag{1.98}$$

For the fluid (1.91), this becomes

$$\rho \geq 0 \qquad \text{and} \qquad \rho + P \geq 0. \tag{1.99}$$

The *dominant energy condition* (DEC) is satisfied if WEC is satisfied and, in addition, $T^{\mu\nu}\, t_\mu$ is a null or timelike vector (*i.e.*, $T_{\alpha\beta}T^\beta{}_\gamma\, t^\alpha\, t^\gamma \leq 0$) for any timelike vector t^μ. For the perfect fluid (1.91), this means that

$$\rho \geq |P| \tag{1.100}$$

(heuristically, this means that the speed at which energy flows in this form of matter does not exceed the speed of light).

The *null energy condition* (NEC) consists of

$$T_{\alpha\beta}\, l^\alpha\, l^\beta \geq 0 \qquad \text{for all null vectors } l^\mu \tag{1.101}$$

which, for the perfect fluid (1.91), becomes

$$\rho + P \geq 0. \tag{1.102}$$

The NEC is violated in cosmology if the expansion of the universe is super-accelerated, *i.e.*, $dH/dt > 0$, where $H(t)$ is the Hubble function (see below). In this case the fluid is called a phantom fluid.

The *null dominant energy condition* (NDEC) consists of

$$T_{\alpha\beta}\, l^\alpha\, l^\beta \geq 0 \text{ and } T^{\alpha\beta}\, l_\beta \text{ is null or timelike for any null vector } l^\mu.$$

This condition is the same as the DEC but here l^μ is a null, instead of timelike, vector. For the perfect fluid (1.91), this means

$$\rho \geq |P| \qquad \text{or} \qquad \rho = -P. \tag{1.103}$$

The *strong energy condition* (SEC) consists of

$$\left(T_{\alpha\beta} - \frac{1}{2}\, T g_{\alpha\beta} \right) t^\alpha\, t^\beta \geq 0 \quad \text{for any timelike vector } t^\mu. \tag{1.104}$$

For the perfect fluid (1.91), this is equivalent to

$$\rho + P \geq 0 \quad \text{and} \quad \rho + 3P \geq 0. \tag{1.105}$$

SEC is violated by a positive cosmological constant, during inflation in the early universe, and by dark energy with $P < -\rho/3$.

1.1.8 *Einstein equations*

The Einstein equations of general relativity relate the curvature[3] of the metric $g_{\mu\nu}$, which is the unknown variable, with the stress–energy tensor of the matter distribution:

$$\mathcal{R}_{\mu\nu} - \frac{1}{2}g_{\mu\nu}\mathcal{R} + \Lambda g_{\mu\nu} = 8\pi G\, T_{\mu\nu}\,, \qquad (1.106)$$

where Λ is the cosmological constant. The latter can be regarded as an effective fluid with stress–energy tensor $T^{(\Lambda)}_{\mu\nu} = -\Lambda g_{\mu\nu}/(8\pi G)$ and energy density and pressure $\rho_\Lambda = -P_\Lambda = \Lambda/(8\pi G)$.

As is clear from the expression (1.85) of the Ricci tensor in terms of the metric tensor and its first and second derivatives, the Einstein equations are of second order in $g_{\mu\nu}$, a feature shared with the Poisson equation of Newtonian gravity (this is not a coincidence because Einstein theory must reduce to Newtonian gravity in the weak–field, slow–motion limit).

Roughly speaking, one can regard the Einstein equations as expressing the fact that the curvature (*i.e.*, the geometry, contained in the left hand side) equals matter, which causes this curvature.

In the absence of matter ($T_{\mu\nu} = 0$) and of cosmological constant, the Einstein equations reduce to

$$\mathcal{R}_{\mu\nu} = 0\,. \qquad (1.107)$$

In cosmology, the universe is never empty, containing at least the cosmological constant Λ and, realistically, some form of matter.

1.2 Cosmology

The universe contains structures, assembled as stars, galaxies, galaxy groups and clusters, and superclusters. However, at some point we stop encountering larger and larger structures. Averaging over a sufficiently large scale a bit larger than 100 megaparsecs, which is small compared with the radius of curvature of the universe, the universe appears homogeneous and isotropic. The precise observations of the cosmic microwave background confirm that, at the time when this radiation stopped interacting with matter and started travelling free, the universe was isotropic in one part in 10^5.

In cosmology, the homogeneity and isotropy of space (on large scales) are taken to form the most basic postulate, known as the Copernican

---

[3]More precisely, the Einstein tensor $G_{\mu\nu}$ built out of contractions of the curvature tensor $\mathcal{R}_{\mu\nu\rho\sigma}$.

Principle [82, 202, 244, 260, 261, 435]. Spatial homogeneity and isotropy are well–defined spacetime symmetries and, in the description of cosmology based on general relativity and on four–dimensional spacetime manifolds [82, 120, 435], the Copernican Principle dictates that the spacetime manifold used to model the universe must be highly symmetric.

Since we can only access a region of the universe comprising the proximity of our past light cone plus a very small region nearby, we are not able to check observationally the validity of the Copernican Principle in most of the universe. Extrapolating the Copernican Principle from the small portion of the universe accessible to us to its entirety is a huge assumption, but it is the only way to proceed. Philosophically, it is equivalent to stating that we do not occupy a privileged position in the universe and that the small region of the universe accessible to us must be typical.

The large-scale picture of the universe that emerges from general relativity has been remarkably successful in describing its present and past and predicting its future with a relatively simple model. The standard Big Bang model explains the universe around us and has contributed greatly to our knowledge and understanding of the expanding cosmos. The standard Big Bang model is based on three observational pillars: the redshift of galaxies, the relative abundances of light elements, and the cosmic microwave background. The latter is interpreted as the fossil radiation left behind by a hot epoch of the primordial universe [117]. Standard physics has successfully predicted its existence, blackbody spectrum, temperature, and the temperature fluctuations present in it caused by primordial matter seeds [117].

In spite of its overall success, certain problems of the Big Bang model cannot be resolved within the model itself unless extremely special initial conditions are advocated, but need new physics. These problems are known as the horizon problem, the flatness problem, the monopole problem, and the origin of density perturbations problem [244, 260, 261]. Currently, the favorite solution to these problems is the paradigm of inflation in the early universe, which has received observational support but is not demonstrated to have occurred.

1.2.1 *Spatial homogeneity and isotropy*

To describe cosmology, we require spatial homogeneity (*i.e.*, all points of three–dimensional space are equivalent) and spatial isotropy (all directions of three–dimensional space are equivalent). These assumptions are very restrictive and imply that one can foliate the four–dimensional spacetime with three–dimensional hypersurfaces Σ_t which are surfaces of constant time t, the proper time of the comoving observers who see the cosmic microwave background homogeneous and isotropic around them (apart from the above–mentioned small temperature fluctuations). The four–dimensional spacetime is composed of these three–dimensional slices Σ_t and of the time direction orthogonal to them (the so-called $3+1$ splitting of spacetime). A $3+1$ spacetime splitting is always relative to a family of observers. While different observers experience different times and three–spaces, because of spatial homogeneity and isotropy there is a preferred family of observers, the comoving observers adapted to the spatial symmetries. As a consequence, there are a preferred notion of time, of three–space, and a preferred $3+1$ splitting because of the symmetries, which makes cosmology a very special case of general relativity.

The time t ("comoving time") is a global time coordinate defined in a geometrically invariant way and the proper time of the comoving observers in whose frame the cosmic fluid filling the universe is at rest.

Let u^μ be the four–velocity field of the family of comoving observers and of the cosmic fluid. In comoving coordinates $\left(t, x^i\right)$ (where the index $i = 1, 2, 3$ runs over the three spatial dimensions), its components are

$$u^\mu = \delta^{\mu 0} = \left(1, 0, 0, 0\right).\tag{1.108}$$

Comoving coordinates are constructed as follows: label each point of a hypersurface Σ_t of constant time t with three coordinates $\left(x^1, x^2, x^3\right)$, then carry these labels to the next hypersurface $\Sigma_{t+\delta t}$ (with $\delta t > 0$) along comoving worldlines (which have u^μ as their tangent) orthogonal to Σ_t. Finally, label each surface Σ_t with the corresponding value of the comoving time t and we have the comoving coordinates $\left(t, x^1, x^2, x^3\right)$.

Since the worldlines are orthogonal to the hypersurfaces Σ_t, the spacetime line element assumes the form

$$ds^2 = -dt^2 + g_{ij} dx^i dx^j\tag{1.109}$$

$(i, j = 1, 2, 3)$, *i.e.*, in these coordinates there are no cross–terms $dt\, dx^i$ in the line element ($g_{0i} = 0$), nor off–diagonal components of the metric tensor.

Since the universe expands, any pair of points on a hypersurface Σ_t move apart from each other by a factor which is independent of the position on Σ_t because of spatial homogeneity. Then, the line element must have the form

$$ds^2 = -dt^2 + a^2(t)\, g_{ij}(x^k)\, dx^i dx^j\,,\tag{1.110}$$

where the "scale factor" $a(t)$ (the same for all directions because of spatial isotropy) is a positive function of time. An increasing [decreasing] $a(t)$ describes an expanding [contracting] universe.

On each hypersurface Σ_t all the spatial points are equivalent, which implies that the curvature of this three–dimensional space cannot depend on the position on Σ_t: the spatial slices Σ_t are three–spaces of constant curvature.[4]

All the spaces of constant curvature (in any dimension) are classified [125]. The Riemann tensor of such a space necessarily has the simple form

$$R_{\alpha\beta\gamma\delta} = K \left(g_{\alpha\gamma}\, g_{\beta\delta} - g_{\alpha\delta}\, g_{\beta\gamma} \right) , \qquad (1.111)$$

where K is the constant curvature with dimensions $\left[K\right] = \left[L^{-2}\right]$. There are only three possible geometries for constant curvature spaces and they differ according to the sign of the constant K.

In dimension three (for the constant time slices of cosmology), the curvature tensor of a space of constant curvature K with metric $g_{ij}^{(3)}$ is simply[5]

$$\mathcal{R}_{ijkl}^{(3)} = K \left(g_{ik}^{(3)} g_{jl}^{(3)} - g_{il}^{(3)} g_{jk}^{(3)} \right) . \qquad (1.112)$$

Contracting with $g_{(3)}^{jl}$, one obtains the three–dimensional Ricci tensor

$$\mathcal{R}_{ik}^{(3)} = g_{(3)}^{jl}\, \mathcal{R}_{ijkl}^{(3)} = K \left(\underbrace{g_{(3)}^{jl} g_{jl}^{(3)}}_{3} g_{ik}^{(3)} - \underbrace{g_{(3)}^{jl} g_{il}^{(3)}}_{\delta_i^j} g_{jk}^{(3)} \right)$$

$$= K \left(3\, g_{ik}^{(3)} - g_{ik}^{(3)} \right)$$

$$= 2K\, g_{ik}^{(3)} , \qquad (1.113)$$

and the three–dimensional Ricci scalar

$$\mathcal{R}^{(3)} = 6K . \qquad (1.114)$$

Three–space is isotropic, hence spherically symmetric about any of its points and its line element must have the form

$$d\ell_{(3)}^2 = c^{\lambda(r)} dr^2 + r^2 \left(d\vartheta^2 + \sin^2 \vartheta\, d\varphi^2 \right) \equiv e^{\lambda(r)} dr^2 + r^2 d\Omega_{(2)}^2 \qquad (1.115)$$

---

[4]Here we refer to the curvature *of the three–dimensional hypersurfaces* Σ_t and not to the curvature of the four–dimensional spacetime.

[5]No derivatives of the metric appear here while, in general, $\mathcal{R}^{\alpha}{}_{\alpha\beta\gamma}$ involves non-linear combinations its first and second derivatives.

in spherical coordinates $\left(r, \vartheta, \varphi\right)$, where $d\Omega_{(2)}^2 \equiv d\vartheta^2 + \sin^2 \vartheta \, d\varphi^2$ denotes the line element on the unit two–sphere. The only non–vanishing components of the Ricci tensor of the constant curvature three–spaces with metric (1.115) are (see, *e.g.*, [260])

$$\mathcal{R}_{11}^{(3)} = \frac{\lambda'}{r}, \tag{1.116}$$

$$\mathcal{R}_{22}^{(3)} = 1 + e^{-\lambda}\left(\frac{r\,\lambda'}{2} - 1\right) = \frac{\mathcal{R}_{33}^{(3)}}{\sin^2 \vartheta}, \tag{1.117}$$

where a prime denotes differentiation with respect to r. The components $(1,1)$ and $(2,2)$ of Eq. (1.113) then yield the system of first order ordinary differential equations

$$\frac{\lambda'}{r} = 2K\,e^{\lambda}, \tag{1.118}$$

$$1 + e^{-\lambda}\left(\frac{r\,\lambda'}{2} - 1\right) = 2Kr^2. \tag{1.119}$$

Substituting λ' from Eq. (1.118) into Eq. (1.119) gives

$$1 + K\,r^2 - e^{-\lambda} = 2K\,r^2 \tag{1.120}$$

and

$$e^{-\lambda} = 1 - K r^2. \tag{1.121}$$

The line element of a three–space of constant curvature K is therefore given by

$$d\ell_{(3)}^2 = \frac{dr^2}{1 - Kr^2} + r^2 d\Omega_{(2)}^2. \tag{1.122}$$

1.2.2 *Friedmann–Lemaître–Robertson–Walker spacetimes*

Based on the previous conclusion and remembering Eqs. (1.110) and (1.122), the line element of a spatially homogeneous and isotropic universe in comoving coordinates $\left(t, r, \vartheta, \varphi\right)$ is the Friedmann–Lemaître–Robertson–Walker (FLRW) line element

$$ds^2 = -dt^2 + a^2(t)\left[\frac{dr^2}{1 - Kr^2} + r^2\left(d\vartheta^2 + \sin^2 \vartheta \, d\varphi^2\right)\right]. \tag{1.123}$$

Its structure describes in a simple way the fact that the three–geometry (the quantity containing the square bracket in the right hand side) evolves with the cosmic time t. This three–space expands if $a(t)$ increases and contracts if $a(t)$ decreases. In the special case $K = 0$, three–space is spatially flat, while it is curved if $K \neq 0$. In the trivial case $a = $ const. and $K = 0$, spacetime reduces to the Minkowski spacetime of special relativity. However, if $a = $ const. but $K \neq 0$, one has a static curved spacetime.[6]

The FLRW line element can be written in different ways. We can redefine the radial coordinate and the scale factor so that the constant three–curvature K is normalized to ± 1 or 0:

$$K \equiv |K| \, \kappa \qquad \text{with } \kappa = \pm 1, 0 \,, \tag{1.124}$$

$$r_* \equiv \sqrt{|K|} \, r \,, \tag{1.125}$$

$$a_*(t) \equiv \begin{cases} \dfrac{a(t)}{\sqrt{|K|}} & \text{if } K \neq 0 \,, \\[3mm] a(t) & \text{if } K = 0 \,. \end{cases} \tag{1.126}$$

If this redefinition is made then, for $\kappa = \pm 1$, one cannot set the scale factor to unity at the present time t_0, $a(t_0) \equiv a_0 = 1$, as is often done in the literature.

The dimensionless quantity κ, which can only assume the values $0, \pm 1$ is the "curvature index". Using this normalized curvature index, we can write the FLRW line element in comoving coordinates as

$$ds^2 = -dt^2 + a_*^2(t) \left(\frac{dr_*^2}{1 - \kappa \, r_*^2} + r_*^2 \, d\Omega_{(2)}^2 \right) . \tag{1.127}$$

The curvature index κ defined by Eq. (1.124) is dimensionless, which implies that r_* is also dimensionless and the rescaled scale factor a_* carries the dimensions of a length. For $K = 0$ there is freedom to choose either

$$\left[a \right] = \left[L \right] \quad \text{and} \quad \left[r \right] = \left[0 \right] \tag{1.128}$$

or

$$\left[a \right] = \left[0 \right] \quad \text{and} \quad \left[r \right] = \left[L \right] . \tag{1.129}$$

Following standard procedure and for economy of notations, from now on we drop the asterisk from r_* and a_*. If needed, the reader will be reminded whether these quantities are the normalized ones.

---

[6]If the scale factor is constant, it can be set to unity because one can always rescale the spatial coordinates to absorb this constant.

Let us denote with an overdot the derivative with respect to the comoving time t. The Hubble function

$$H(t) \equiv \frac{\dot{a}}{a} \qquad (1.130)$$

has the dimensions of the inverse of a time and H^{-1} is often taken as an approximation to the age of the universe at the time t. The "Hubble radius" H^{-1} gives, in order of magnitude, the radius of curvature of the universe.

1.2.3   Geometry of three–spaces of constant curvature

Consider again the FLRW line element (1.127)

$$ds^2 = -dt^2 + a^2(t)\left(\frac{dr^2}{1 - \kappa r^2} + r^2\, d\Omega^2_{(2)}\right); \qquad (1.131)$$

every slice Σ_t of constant comoving time t has three–geometry given by

$$ds^2\Big|_{\Sigma_t} = a^2(t)\left(\frac{dr^2}{1 - \kappa r^2} + r^2 d\Omega^2_{(2)}\right) \equiv a^2(t)\, d\ell^2_{(3)}. \qquad (1.132)$$

Let us consider separately the three possible geometries corresponding to the sign of the curvature index κ.

1.2.3.1   Three–sphere (closed universe)

If $\kappa = +1$ the fraction $1/(1 - \kappa r^2)$ diverges as $r \to 1$ (the coordinate r, originally denoted with r_*, is dimensionless). Let us introduce the new dimensionless radial coordinate χ (hyperspherical radius) defined by $r \equiv \sin\chi$ and spanning the range $0 \le \chi < \pi$: we have

$$dr = \cos\chi\, d\chi = \sqrt{1 - \sin^2\chi}\, d\chi = \sqrt{1 - r^2}\, d\chi \qquad (1.133)$$

and the three–dimensional line element (1.132) becomes

$$ds^2\Big|_{\Sigma_t} = a^2(t)\left(d\chi^2 + \sin^2\chi\, d\Omega^2_{(2)}\right). \qquad (1.134)$$

This metric describes the geometry of a three–space that can be embedded in the four–dimensional Euclidean space $\left(w, x, y, z\right)$ as

$$w = a \cos \chi \,, \tag{1.135}$$

$$x = a \sin \chi \sin \vartheta \cos \varphi \,, \tag{1.136}$$

$$y = a \sin \chi \sin \vartheta \sin \varphi \,, \tag{1.137}$$

$$z = a \sin \chi \cos \vartheta \,. \tag{1.138}$$

This embedding is possible because

$$dw^2 + dx^2 + dy^2 + dz^2 = a^2 \left(d\chi^2 + \sin^2 \chi \, d\Omega_{(2)}^2 \right) \,. \tag{1.139}$$

The equation of the three–space embedded in the four–dimensional Euclidean space $\left(w, x, y, z\right)$ is

$$w^2 + x^2 + y^2 + z^2 = a^2 \,, \tag{1.140}$$

that is, this space with constant positive curvature is a three–sphere with finite volume

$$V(t) = 2\pi^2 a^3 (t) \,. \tag{1.141}$$

In fact, using the line element on constant time slices Σ_t

$$ds^2 \Big|_{\Sigma_t} = a^2 (t) \left(d\chi^2 + \sin^2 \chi \, d\Omega_{(2)}^2 \right) \,, \tag{1.142}$$

the volume of the universe (that is, the volume of the three–space Σ_t) is

$$V = \int_0^\pi d\chi \int_0^\pi d\vartheta \int_0^{2\pi} d\varphi \, \sqrt{g^{(3)}} \,, \tag{1.143}$$

where $g^{(3)} = a^6 \sin^4 \chi \sin^2 \vartheta$ is the determinant of the three–metric $g_{ij}^{(3)}$ on Σ_t. One obtains

$$V = a^3 (t) \int_0^\pi d\chi \sin^2 \chi \int_0^\pi d\vartheta \sin \vartheta \cdot 2\pi = 2\pi^2 a^3 \,,$$

where we used the elementary integrals

$$\int_0^\pi d\chi \sin^2 \chi = \frac{\pi}{2} \,, \qquad \int_0^\pi d\vartheta \sin \vartheta = 2 \,. \tag{1.144}$$

Embedding the three–space of constant positive curvature into a fictitious four–dimensional Euclidean space is a mathematical trick without physical meaning.

The three–sphere has finite volume, it has no boundary, and its topology is closed and compact, hence the corresponding universe is finite with time–dependent volume $V(t) = 2\pi^2 a^3(t)$. An observer in this universe travelling along a geodesic curve eventually returns to its starting point. The topology of the four–dimensional universe is $\mathbb{R} \times S^3$, where $\mathbb{R}$ denotes the time direction and each slice Σ_t is a three–sphere S^3.

1.2.3.2   *Flat space*

If $\kappa = 0$ the spatial sections Σ_t are Euclidean. This universe is known as the critically open or spatially flat universe. We can use the polar coordinates $\left(r, \vartheta, \varphi\right)$ or the Cartesian coordinates

$$x = r \sin \vartheta \cos \varphi \,, \tag{1.145}$$

$$y = r \sin \vartheta \sin \varphi \,, \tag{1.146}$$

$$z = r \cos \vartheta \,, \tag{1.147}$$

to write the three–dimensional Euclidean line element of each slice Σ_t as

$$ds^2 \Big|_{\Sigma_t} = a^2(t) \left(dx^2 + dy^2 + dz^2 \right) \tag{1.148}$$

$$= a^2(t) \left(dr^2 + r^2 d\Omega_{(2)}^2 \right) \,. \tag{1.149}$$

The coordinates span the ranges

$$-\infty < x < +\infty, \quad -\infty < y < +\infty, \quad -\infty < z < +\infty, \tag{1.150}$$

$$0 < r < +\infty, \quad 0 \le \vartheta < \pi, \quad 0 \le \varphi < 2\pi \,. \tag{1.151}$$

The three–dimensional space is the Euclidean $\mathbb{R}^3$ and the topology of the four–dimensional spacetime is $\mathbb{R} \times \mathbb{R}^3$, with the Lorentzian spacetime metric $g_{\mu\nu}$.

1.2.3.3 *Hyperbolic space (open universe)*

The $\kappa = -1$ case corresponds to the hyperbolic three–space and to the open universe. The spacetime line element in comoving coordinates is

$$ds^2 = -dt^2 + a^2(t) \left(\frac{dr^2}{1 + r^2} + r^2 d\Omega_{(2)}^2 \right) . \tag{1.152}$$

We introduce the new dimensionless radial coordinate χ (hyperspherical radius) defined by $r \equiv \sinh \chi$ spanning the range $0 \leq \chi < +\infty$. The differentials of r and χ are related by $dr = \cosh \chi \, d\chi$ and the line element on the three–slices Σ_t of constant time is

$$ds^2 \Big|_{\Sigma_t} = a^2(t) \left(\frac{\cosh^2 \chi}{1 + \sinh^2 \chi} d\chi^2 + \sinh^2 \chi \, d\Omega_{(2)}^2 \right)$$

$$= a^2(t) \left(d\chi^2 + \sinh^2 \chi \, d\Omega_{(2)}^2 \right) . \tag{1.153}$$

One can not embed this three–surface into Euclidean four–space, however it is possible to embed it into a fictitious four–dimensional Minkowski space $\left(w, x, y, z \right)$ with line element

$$d\sigma^2 = -dw^2 + dx^2 + dy^2 + dz^2 \tag{1.154}$$

by setting

$$w = a \cosh \chi , \tag{1.155}$$

$$x = a \sinh \chi \sin \vartheta \cos \varphi , \tag{1.156}$$

$$y = a \sinh \chi \sin \vartheta \sin \varphi , \tag{1.157}$$

$$z = a \sinh \chi \cos \vartheta . \tag{1.158}$$

Then, in four–dimensional Minkowski spacetime it is

$$d\sigma^2 = -dw^2 + dx^2 + dy^2 + dz^2 = a^2 \left(d\chi^2 + \sinh^2 \chi \, d\Omega_{(2)}^2 \right) . \tag{1.159}$$

The hyperbolic three–space of negative curvature is described by the equation

$$w^2 - x^2 - y^2 - z^2 = a^2 , \tag{1.160}$$

that is, by the equation of a three–dimensional hyperboloid in the four–dimensional Minkowski spacetime. The spherical coordinates span the ranges

$$0 \leq \chi < +\infty , \quad 0 \leq \vartheta < \pi , \quad 0 \leq \varphi < 2\pi . \tag{1.161}$$

The two–surfaces $\chi = $ const. are two–spheres with metric $g^{(2)}_{\mu\nu}$ and area

$$A(t,\chi) = \int_0^\pi d\vartheta \int_0^{2\pi} d\varphi \sqrt{g^{(2)}}$$

$$= \int_0^\pi a \sinh \chi \sin \vartheta \, d\vartheta \int_0^{2\pi} a \sinh \chi \, d\varphi$$

$$= 4\pi a^2(t) \sinh^2 \chi. \tag{1.162}$$

For small radial distances $\chi \to 0$, this three–space is well approximated by Euclidean space, $\sinh \chi \simeq \chi$, and $A \simeq 4\pi a^2 \chi^2 + \dots$ (this fact corresponds to the general idea of differential geometry that a curved manifold is approximated locally by its tangent space). If instead χ is large, the area A is proportional to $\sinh^2 \chi$ and increases faster than in Euclidean space (in which it is instead $A \propto \chi^2$). The volume of the universe (meaning the volume of the three–sections Σ_t) is infinite:

$$V = 4\pi a^2 \int_0^{+\infty} d\chi \sinh^2 \chi = +\infty. \tag{1.163}$$

To summarize, the FLRW line element with normalized curvature index κ can be written in comoving coordinates as

$$ds^2 = -dt^2 + a^2(t) \left[d\chi^2 + f^2(\chi) \left(d\vartheta^2 + \sin^2 \vartheta \, d\varphi^2 \right) \right], \tag{1.164}$$

where

$$f(\chi) = r = \begin{cases} \sin \chi & \text{if} \quad \kappa = +1, \\[2mm] \chi & \text{if} \quad \kappa = 0, \\[2mm] \sinh \chi & \text{if} \quad \kappa = -1, \end{cases} \tag{1.165}$$

and where χ is a dimensionless radial coordinate (hyperspherical radius), while a carries the dimensions of a length. In the spatially flat ($\kappa = 0$) case one can give χ the dimensions of a length and keep the scale factor a dimensionless. In all cases the proper (physical) radial distance is $a(t)\chi$.

The high degree of symmetry imposed by requiring that the universe be spatially homogeneous and isotropic has reduced the freedom in the geometry to a single unknown function of a single variable, the scale factor $a(t)$, plus the three possibilities $\kappa = 0, \pm 1$ (the choice of the latter, however, is not part of the dynamics[7]). A consequence of this very significant restriction is that[8] all FLRW metrics are conformally flat: a direct computation shows that the Weyl tensor vanishes identically,

$$C_{\mu\nu\rho}{}^{\sigma} = 0 . \tag{1.166}$$

It is almost trivial to realize the conformal flatness of FLRW spacetime in the spatially flat ($\kappa = 0$) case. In fact, using the conformal time η defined by

$$dt \equiv a \, d\eta , \tag{1.167}$$

the $\kappa = 0$ FLRW line element is written as

$$ds^2 = a^2(\eta)\left(-d\eta^2 + dx^2 + dy^2 + dz^2 \right), \tag{1.168}$$

which is clearly conformal to the Minkowski metric with conformal factor equal to the scale factor, $\Omega = a(\eta)$. In the $\kappa = \pm 1$ cases, in order to bring the line element into a form that shows explicitly the conformal flatness one needs to perform more involved coordinate transformations [120]. In any case, a direct computation of the Weyl tensor $C_{\alpha\beta\gamma}{}^{\delta}$ using the FLRW line element shows that this quantity vanishes for any value of the curvature index κ [435].

In the FLRW line element

$$ds^2 = -dt^2 + a^2(t) \, d\ell_{(3)}^2 = -dt^2 + a^2(t) \left(d\chi^2 + f^2(\chi) \, d\Omega_{(2)}^2 \right) \tag{1.169}$$

the comoving coordinates are adjusted to follow the expansion of the universe and do not correspond to the physical notion of distance between two points. The coordinate distance or comoving distance between two points 1 and 2 is

$$\ell_{12} = \int_1^2 d\ell_{(3)} . \tag{1.170}$$

The proper distance or physical distance between two points

$$d_{\text{phys}}(t) = a(t) \, \ell_{12} \tag{1.171}$$

is instead the comoving distance multiplied by the scale factor and depends on time. In an expanding universe all physical lengths do the same.

---

[7]The topology of the spatial sections Σ_t cannot change with time: a spherical universe remains spherical, a spatially flat one remains spatially flat, *etc.*

[8]The spacetime metric $\tilde{g}_{\mu\nu}$ is said to be conformal to the metric $g_{\mu\nu}$ if there exists a nowhere–vanishing regular function $\Omega(x^\alpha)$ of the spacetime position such that $\tilde{g}_{\mu\nu} = \Omega^2 g_{\mu\nu}$. A conformal transformation preserves angles, the timelike/null/spacelike character of four–vectors, and the causal structure of spacetime [435].

This property characterizes all physical lengths in FLRW space, therefore also the physical wavelengths of waves propagating in FLRW spacetimes. If λ is the comoving wavelength of a wave, the physical wavelength is $\lambda_{\text{phys}} = a(t)\,\lambda$. A wave that propagates in an expanding FLRW space is stretched by its expansion. This stretching of wavelengths applies to light propagating to us from sources at cosmological distances and is called cosmological redshift.

1.3   Einstein–Friedmann equations

The form of matter filling the universe must be assumed with physical insight and studied with astronomical observations or local experiments. In cosmology it is usually assumed that, during a given epoch of the history of the universe, its matter content is described by a single perfect fluid which we refer to as the cosmic fluid and has stress–energy tensor $T_{\mu\nu}$ of the form (1.91). u^{μ} coincides with the four–velocity of comoving observers, while the energy density and pressure depend only on the comoving time, $\rho = \rho(t)$ and $P = P(t)$. In comoving coordinates, the contravariant components of the fluid four–velocity are $u^{\mu} = \delta^{\mu 0} = \left(1, 0, 0, 0\right)$ and its covariant components are $u_{\mu} = g_{\mu\nu} u^{\nu} = g_{\mu\nu} \delta^{\nu 0} = g_{00}\,\delta_{\mu 0} = \left(-1, 0, 0, 0\right)$.

The components of the stress–energy tensor of the cosmic fluid in comoving coordinates are then $T_{\mu\nu} = (P + \rho)\,\delta_{\mu 0}\delta_{\nu 0} + Pg_{\mu\nu}$, which gives

$$T_{00} = P + \rho + Pg_{00} = \rho\,,$$

$$T_{0i} = (P + \rho)\,\delta_{00}\,\delta_{0i} + Pg_{0i} = 0 \qquad (i = 1, 2, 3),$$

$$T_{11} = (P + \rho)\,\delta_{10}\,\delta_{10} + Pg_{11} = \frac{Pa^2}{1 - \kappa\,r^2}\,,$$

$$T_{22} = (P + \rho)\,\delta_{20}\,\delta_{20} + Pg_{22} = Pa^2 r^2\,,$$

$$T_{33} = (P + \rho)\,\delta_{30}\,\delta_{30} + Pg_{33} = Pa^2 r^2 \sin^2\theta\,,$$

while, for the spatial indices $i \neq j$, it is

$$T_{ij} = (P + \rho)\,\delta_{i0}\,\delta_{j0} + Pg_{ij} = 0\,. \qquad (1.172)$$

In matrix form, we have

$$(T_{\mu\nu}) = \begin{pmatrix} \rho & 0 & 0 & 0 \\ 0 & \frac{Pa^2}{1-\kappa r^2} & 0 & 0 \\ 0 & 0 & Pa^2 r^2 & 0 \\ 0 & 0 & 0 & Pa^2 r^2 \sin^2\theta \end{pmatrix}. \tag{1.173}$$

In the spatially flat case $\kappa = 0$ one can use Cartesian coordinates $\left(t, x, y, z\right)$, in terms of which the line element is

$$ds^2 = -dt^2 + a^2(t)\left(dx^2 + dy^2 + dz^2\right) = -dt^2 + a^2(t)\,\delta_{ij} dx^i dx^j \tag{1.174}$$

and Eq. (1.91) yields

$$T_{11} = (P + \rho)\,\delta_{10}\,\delta_{10} + Pg_{11} = Pa^2\,,$$

$$T_{22} = (P + \rho)\,\delta_{20}\,\delta_{20} + Pg_{22} = Pa^2\,,$$

$$T_{33} = (P + \rho)\,\delta_{30}\,\delta_{30} + Pg_{33} = Pa^2\,,$$

and

$$(T_{\mu\nu}) = \begin{pmatrix} \rho & 0 & 0 & 0 \\ 0 & Pa^2 & 0 & 0 \\ 0 & 0 & Pa^2 & 0 \\ 0 & 0 & 0 & Pa^2 \end{pmatrix} \tag{1.175}$$

which, of course, agrees with Eq. (1.173) with $\kappa = 0$. Given the FLRW line element (1.127) we compute the Ricci tensor $R_{\mu\nu}$, the Ricci scalar R, and the Einstein tensor $G_{\mu\nu}$ and we write the Einstein equations with the perfect fluid stress–energy tensor (1.91) in the right hand side. This computation is reported in Appendix A.1 for the spatially flat case. The time–time component of the Einstein equations (called Hamiltonian or scalar constraint in general relativity) gives the Friedmann equation

$$H^2 = \frac{8\pi G}{3}\,\rho\,. \tag{1.176}$$

This is a first order differential equation for the scale factor $a(t)$ and, therefore, a constraint on the dynamics and not a fully dynamical second order equation.

The $(1,1)$ component of the Einstein equations gives

$$-\frac{2\,\ddot{a}}{a} - \frac{\dot{a}^2}{a^2}\,\rho = 8\pi GP\,. \tag{1.177}$$

Substituting Eq. (1.176) into this one yields the "acceleration equation"

$$\frac{\ddot{a}}{a} = -\frac{4\pi G}{3}\left(\rho + 3P\right).\qquad(1.178)$$

Due to the spatial isotropy, the $(2,2)$ and $(3,3)$ components of the Einstein equations coincide with the $(1,1)$ component.

1.3.1 *General case*

The calculations for spatially curved universes with $\kappa = \pm 1$ are longer due to the spatial curvature and to the use of polar, instead of Cartesian, coordinates. The results for general $\kappa = 0, \pm 1$ give the Einstein–Friedmann equations of FLRW cosmology with a perfect fluid[9] [255]

$$H^2 = \frac{8\pi G}{3}\rho - \frac{\kappa}{a^2} + \frac{\Lambda}{3},\qquad(1.179)$$

$$\frac{\ddot{a}}{a} = -\frac{4\pi G}{3}\left(\rho + 3P\right) + \frac{\Lambda}{3}.\qquad(1.180)$$

This is a set of two ordinary differential equations for the scale factor $a(t)$, with the Friedmann equation providing a first order constraint on the dynamics. These are the fundamental equations of FLRW cosmology. Contrary to Newtonian gravity, one cannot prescribe *a priori* the dynamics of the matter fluid described, but one must solve the Einstein equations simultaneously for the scale factor $a(t)$ and for the matter variables $\rho(t)$ and $P(t)$.

The expression of the Ricci scalar in general FLRW spaces

$$\mathcal{R} = 6\left(\dot{H} + 2H^2 + \frac{\kappa}{a^2}\right) = 6\left(\frac{\ddot{a}}{a} + \frac{\dot{a}^2}{a^2} + \frac{\kappa}{a^2}\right)\qquad(1.181)$$

is useful in practical calculations.

According to the acceleration equation (1.180), if the strong energy condition $(\rho + 3P \geq 0)$ is satisfied, it is $\ddot{a} \leq 0$ and the universe cannot accelerate. Let us discuss open and closed universes separately.

---

[9]The Einstein–Friedmann equations (1.179) and (1.180) were derived by Georges Lemaître, who realized their relevance in connection with the discovery of the cosmic expansion by Hubble [255] and should, therefore, be called Lemaître equations, although this would not be standard terminology.

1.3.1.1 *Open universe ($\kappa = -1$)*

In this case the strong energy condition, if imposed, implies that $\rho \geq 0$ and $\ddot{a} \leq 0$. The Friedmann equation (with $\Lambda = 0$)

$$H^2 = \frac{8\pi G}{3}\rho + \frac{1}{a^2} \tag{1.182}$$

with $\rho \geq 0$ guarantees that $\dot{a} \neq 0$ or, these observers cannot see the universe as static (but if the universe is empty, $\rho = 0$, it is indeed static—Minkowski spacetime in accelerated coordinates, called Milne universe [288]). Under the strong energy condition, there are no static open non–empty FLRW universes. Therefore, the universe must decelerate if the cosmic fluid satisfies $\rho + 3P > 0$.

1.3.1.2 *Closed universe ($\kappa = +1$)*

In this case the strong energy condition still implies $\ddot{a} \leq 0$. The Friedmann equation is now

$$H^2 = \frac{8\pi G}{3}\rho - \frac{1}{a^2} \tag{1.183}$$

and, in order to keep $H^2 \geq 0$, it must be $a \leq a_{\max}$, where

$$a_{\max} = \sqrt{\frac{3}{8\pi G\rho_{\min}}}, \tag{1.184}$$

and $\rho_{\min}$ is the minimum density: this universe must have finite size. If it is expanding, it can only expand up to a maximum value of the scale factor and a maximum size but no further, consistent with the fact that the three–space Σ_t has always finite volume.

Let us look for static solutions. It can be $\dot{a} \equiv 0$ if and only if $a = a_s$. In particular, the universe can be static if and only if the density stays constant ($\rho = \rho_s$) and $a(t) = a_s$ for all times: this solution corresponds to the "Einstein static universe" which is discussed later.

1.3.2 *Energy conservation*

By definition, the universe comprises all that exists and, therefore, it cannot exchange energy with its surroundings. Then, the energy of the cosmic fluid is conserved in an appropriate (covariant) sense. Differentiate the Friedmann equation (1.179) with respect to time, obtaining

$$\frac{2\dot{a}\ddot{a}}{a^2} - \frac{2\dot{a}^3}{a^3} = \frac{8\pi G}{3}\dot{\rho} + 2\kappa \frac{\dot{a}}{a^3} \tag{1.185}$$

and use the acceleration equation (1.180) to substitute for $\ddot{a}/a$, which gives

$$- H\left(3P + \rho\right) - 2H\left(H^2 + \frac{\kappa}{a^2}\right)\frac{3}{8\pi G} = \dot{\rho} ;\qquad (1.186)$$

use again the Friedmann equation (1.179) to substitute for $H^2 + \kappa/a^2$, obtaining the first order "conservation equation"

$$\dot{\rho} + 3H\left(P + \rho\right) = 0 .\qquad (1.187)$$

This conservation equation is derived from the other two Einstein–Friedmann equations and is not independent, but it often provides a more convenient way of looking at the physics and the dynamics. It can be derived directly from $\nabla^\nu T_{\mu\nu} = 0$ (which is a consequence of the contracted Bianchi identities $\nabla^\nu G_{\mu\nu} = 0$ and of the Einstein equations[10]).

Equation (1.187) has a thermodynamic interpretation. Consider a unit comoving volume and the corresponding physical volume $V = a^3$, then Eq. (1.187) can be rewritten as

$$\frac{d}{dt}\left(\rho a^3\right) + P\frac{d\left(a^3\right)}{dt} = 0 .\qquad (1.188)$$

The first term of Eq. (1.188) can be interpreted as the rate of change of the internal energy $U = \rho a^3$ contained in the physical volume $V = a^3$. In the comoving time dt, the changes of internal energy U and volume V are related by

$$dU + PdV = 0 ,\qquad (1.189)$$

which is the familiar form of the first law of thermodynamics $dQ = dU + PdV$ when the heat exchanged dQ vanishes. Since we are considering a comoving volume, Eq. (1.189) expresses the fact that there can be no energy exchange between neighbouring comoving volumes, otherwise there would be spatial energy currents breaking spatial isotropy.

1.3.2.1   *Evolution of the Hubble function*

Differentiating the Friedmann equation (1.179) and substituting in the result the conservation equation (1.187) yields the rate of change of the Hubble function

$$\dot{H} = -4\pi G\left(P + \rho\right) + \frac{\kappa}{a^2} .\qquad (1.190)$$

---

[10]There is a deeper meaning to the equation $\nabla^\nu T_{\mu\nu} = 0$ for the cosmic fluid: it is a consequence of the invariance of the matter part of the action under spacetime diffeomorphisms [435], which means that it does not require assumptions on the field equations and the theory of gravity.

Equation (1.190) is sometimes called a propagation equation. The Hubble function is related to the expansion scalar Θ of the congruence of timelike worldlines [435] of comoving observers by $\Theta = 3H$ and Eq. (1.190) can be seen as describing the evolution of Θ along these worldlines.[11] Although useful, this equation is derived using only the Einstein–Friedmann equations and does not contain new information.

1.3.3 Integration of the energy equation for constant equation of state

If the cosmic fluid has a constant barotropic equation of state

$$P = w\rho, \qquad w = \text{constant}, \tag{1.191}$$

Eq. (1.187) is immediately integrated giving ρ, and then P, as functions of the scale factor a. Using the equation of state (1.191) in the energy conservation equation gives

$$\dot{\rho} + 3(w+1)H\rho = 0, \tag{1.192}$$

which has the solution

$$\rho(a) = \frac{\rho_*}{a^{3(w+1)}}, \tag{1.193}$$

$$P(a) = w\,\rho(a) = \frac{P_*}{a^{3(w+1)}}, \tag{1.194}$$

if $w \neq -1$, where ρ_* is a constant and $P_* = w\rho_*$.

The value $w = -1$ of the constant equation of state parameter plays a critical role in Eq. (1.193). If $w > -1$, the energy density of the barotropic cosmic fluid decreases with a: this "ordinary" cosmic fluid dilutes with the expansion. If $w < -1$, the fluid becomes more concentrated as the universe expands. The critical value $w = -1$ separating these two types of behaviours corresponds to the cosmological constant, which has constant energy density and does not dilute with the cosmic expansion.[12]

Dust with $P = 0$ has energy density scaling as

$$\rho_{\mathrm{m}}(a) = \frac{\rho_*}{a^3} \tag{1.195}$$

---

[11]In the more general context of general relativity, the equation describing how the expansion Θ evolves as one moves along a worldline is called Raychaudhuri equation and involves extra terms in the right hand side [435].

[12]The curvature index κ does not appear in the conservation equation (1.187) and its solution (1.193) holds for all values of κ.

and corresponds to the reduced form $dU = d\left(\rho a^3\right) = 0$ of the first law of thermodynamics (1.189). The total energy of dust is just the constant rest mass of the non–relativistic fluid particles; while the physical volume scales as a^3, $\rho_{\mathrm{m}} \sim a^{-3}$.

The energy density of radiation in thermal equilibrium, with $w = 1/3$ and $T = -\rho + 3P = 0$, scales as

$$\rho_{\mathrm{rad}}(a) = \frac{\rho_*}{a^4}, \tag{1.196}$$

which is interpreted as follows. Photons have energy $E = h\nu$ (the Planck constant times the frequency ν) which is redshifted by the cosmic expansion since the physical wavelength $\lambda = c/\nu$ scales as the scale factor a, so $E \sim hc/\lambda \sim 1/a$. In addition, the physical volume scales as $V \sim a^3$ and the energy density as a^{-4}.

For the cosmological constant with $w = -1$, the conservation equation (1.187) gives $\dot{\rho} = 0$ and $\rho_\Lambda = \mathrm{const}$.

Phantom energy with $w < -1$ (if it exists) has the peculiar property that $\dot{\rho} = -3(w + 1)H\rho > 0$ in an expanding universe and its energy density scales as

$$\rho_{\mathrm{phantom}} = \rho_* \, a^{3|w+1|}, \tag{1.197}$$

increasing as the universe expands. A phantom fluid causes a Big Rip singularity at a finite time in the future, violating the null energy condition.

Stiff matter with $P = \rho$ has energy density scaling as

$$\rho_{\mathrm{stiff}}(a) = \frac{\rho_*}{a^6}. \tag{1.198}$$

1.3.4   *Dominance of a perfect fluid*

In practice, the matter content of the universe is not a single fluid but several fluids coexist. Usually one of these fluids has energy density much larger than the energy density of the other matter components because, as the universe expands, the energy densities of different fluids scale with different powers of the scale factor, as illustrated in Fig. 1.1.

In standard Big Bang cosmology only non–relativistic matter, radiation, and possibly a cosmological constant are present. In the early universe $a \to 0$, radiation dominates because its energy density scales as $1/a^4$, while the energy density of non–relativistic matter scales only as $1/a^3$ and ρ_Λ remains constant. The curve representing radiation lies above the corresponding curves for dust and Λ as $a \to 0$.

1.4 Perfect fluid solutions

1.4.1 *Spatially flat, $\Lambda = 0$, FLRW universes*

When the cosmic content consists of a single perfect fluid with equation of state $P = w\rho$ with $w = $ const., the energy conservation equation (1.187) immediately integrates to $\rho(a) = \rho_* a^{-3(w+1)}$. This expression can now be substituted into the Friedmann equation for a $\kappa = 0$ FLRW universe

$$H^2 = \frac{8\pi G}{3}\rho \tag{1.199}$$

to obtain

$$a^{\frac{1+3w}{2}}\dot{a} = \sqrt{C}, \tag{1.200}$$

where $C = 8\pi G\rho_*/3$ is constant, which is integrated to

$$a(t) = a_*\left(t - t_*\right)^{\frac{2}{3(w+1)}} \tag{1.201}$$

if $w \neq -1$.

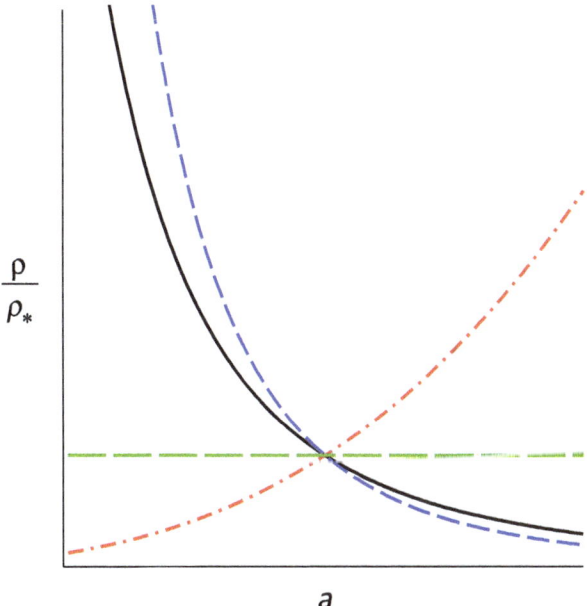

Legend

—— dust   – – radiation   – – – cosmological constant   –·– phantom energy

FIG. 1.1. The densities $\rho(a)$ of dust, radiation, cosmological constant and phantom energy with $w = -2$, as functions of the scale factor a.

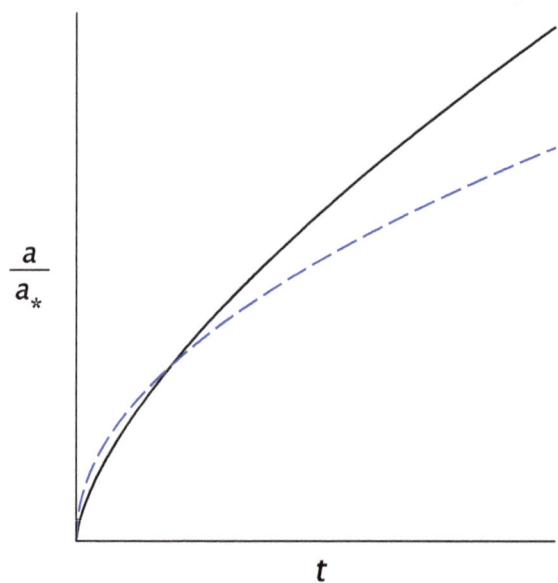

Legend

—— **dust** —— radiation

FIG. 1.2. The scale factor of a spatially flat universe dominated by dust or by radiation.

1.4.1.1   *Dust*

For dust with $w = 0$ the solution for the scale factor is [173]

$$a(t) = a_* \, t^{2/3} \,, \tag{1.202}$$

and the corresponding universe is called "Einstein–de Sitter model" (Fig. 1.2).

When this solution is extrapolated to zero time, as $t \to 0^+$, the scale factor goes to zero and the energy density $\rho(t)$ diverges. The Ricci scalar (an invariant of the curvature tensor)

$$\mathcal{R} = 6 \left(\dot{H} + 2H^2 \right) = \frac{4}{3t^2} \tag{1.203}$$

diverges as $t \to 0$, signalling a spacetime singularity known as the Big Bang. This universe has a beginning at the Big Bang singularity and expands forever afterwards.

1.4.1.2 *Radiation*

With the radiation equation of state $w = 1/3$, the scale factor is

$$a(t) = a_* \sqrt{t}. \tag{1.204}$$

Also in this case there is a Big Bang as $a \to 0^+$ (Fig. 1.2).

1.4.2 *Cosmological constant*

For a pure cosmological constant and no matter, the solution of Eq. (5.69) with $w = -1$ is

$$a(t) = a_* \, e^{Ht} \tag{1.205}$$

with constant Hubble parameter H: this is the exponentially expanding de Sitter universe. The Friedmann equation $H^2 = 8\pi G\rho/3$ implies that the cosmological constant cannot be negative and that the constant Hubble parameter is

$$H = \sqrt{\frac{\Lambda}{3}} \quad (\Lambda \geq 0). \tag{1.206}$$

The de Sitter space is a space of constant (four–dimensional) spacetime curvature. It is possible to recast the de Sitter line element in static form by introducing appropriate (non–comoving) coordinates, but only a portion of the spacetime manifold is covered by these static coordinates [138]. The de Sitter space does not have a beginning like the Einstein–de Sitter model or the radiation–dominated universe: the Big Bang is pushed to $t \to -\infty$. Its importance lies in the fact that it is an attractor in inflationary models of the early universe [193, 244, 261, 264, 384] and in dark energy–dominated models of the late (present–day) accelerating universe [5].

   Let us consider now the case of a negative cosmological constant $\Lambda < 0$. In this case the Friedmann equation

$$H^2 = \frac{\Lambda}{3} - \frac{\kappa}{a^2} \tag{1.207}$$

informs us that, in order to keep H^2 non–negative, there can be solutions only if $\kappa < 0$. Setting $\kappa = -1$, the Friedmann equation can be integrated explicitly obtaining

$$a(t) = \sqrt{\frac{3}{|\Lambda|}} \, \cos\left[\sqrt{\frac{|\Lambda|}{3}} \, (t - t_*)\right], \tag{1.208}$$

corresponding to the "Anti–de Sitter" (or AdS) universe. The cosmological observations tell us that, if the FLRW model is correct, we live in a universe with positive cosmological constant, nevertheless the Anti–de Sitter space is interesting from the mathematical point of view [202]. It has a boundary, it is a constant curvature space and, unlike de Sitter space, it is covered entirely by a static coordinate patch. AdS space is fundamental in string theories because of the AdS/CFT correspondence [220], which posits that physics on the bulk of five–dimensional AdS space is equivalent to a conformal field theory on the four–dimensional AdS boundary. This equivalence has been conjectured to hold in a much broader context, implying that the universe we live in would be a sort of hologram [220].

1.4.2.1   *Solutions for spatially curved universes with $\Lambda = 0$*

In general, for $\kappa = \pm 1$ and constant barotropic equation of state $P = w\rho$ of the cosmic fluid, one cannot obtain an explicit expression $a(t)$ for the scale factor as a function of the comoving time, even for simple equations of state. In many cases, an analytical solution can be written in parametric form using the conformal time η defined by Eq. (1.167) as a parameter.

Choosing the Big Bang as the origin of the time axis and the initial condition $a(0) = 0$, the parametric solution for closed universes is

$$a(\eta) = a_* \left[\sin(\alpha \eta) \right]^{1/\alpha}, \tag{1.209}$$

$$t(\eta) = a_* \int_0^\eta d\eta' \left[\sin(\alpha \eta') \right]^{1/\alpha}, \tag{1.210}$$

$$\alpha = \frac{3w + 1}{2} \qquad (\kappa = +1). \tag{1.211}$$

For open universes, the solution is instead

$$a(\eta) = a_* \left[\sinh(\alpha \eta) \right]^{1/\alpha}, \tag{1.212}$$

$$t(\eta) = a_* \int_0^\eta d\eta' \left[\sinh(\alpha \eta') \right]^{1/\alpha}, \tag{1.213}$$

$$\alpha = \frac{3w + 1}{2} \qquad (\kappa = -1). \tag{1.214}$$

These general formulas contain the following specific cases.

Dust and $\kappa = +1$.

For dust $(w = 0)$ and curvature index $\kappa = +1$, one has [173]

$$a(\eta) = \frac{C}{2}\left(1 - \cos\eta\right),$$
(1.215)

$$t(\eta) = \frac{C}{2}\left(\eta - \sin\eta\right),$$
(1.216)

where

$$C = \frac{8\pi G\rho_*}{3}$$
(1.217)

and $\rho = \rho_*/a^3$.

Radiation and $\kappa = +1$.

For radiation $(P = \rho/3)$ and curvature index $\kappa = +1$, one has

$$a(t) = \sqrt{C}\left[1 - \left(1 - \frac{t}{\sqrt{C}}\right)^2\right]^{1/2},$$
(1.218)

where

$$C = \frac{8\pi G\rho_*}{3}$$
(1.219)

and $\rho = \rho_*/a^4$. In this case there is no need to resort to a parametric representation.

Dust and $\kappa = -1$

For dust $(P = 0)$ and curvature index $\kappa = -1$, one has [173]

$$a(\eta) = \frac{C}{2}\left(\cosh\eta - 1\right),$$
(1.220)

$$t(\eta) = \frac{C}{2}\left(\sinh\eta - \eta\right),$$
(1.221)

with C and ρ as in the $\kappa = +1$ case.

Radiation and $\kappa = -1$

For radiation $(P = \rho/3)$ and curvature index $\kappa = -1$, one has the simple expression

$$a(t) = \sqrt{C}\left[\left(1 - \frac{t}{\sqrt{C}}\right)^2 - 1\right]^{1/2},$$
(1.222)

with C and ρ as in the $\kappa = +1$ case.

For comparison, we report the exact solutions for radiation obtained for the three possible values of the curvature index κ in Fig. 1.3.

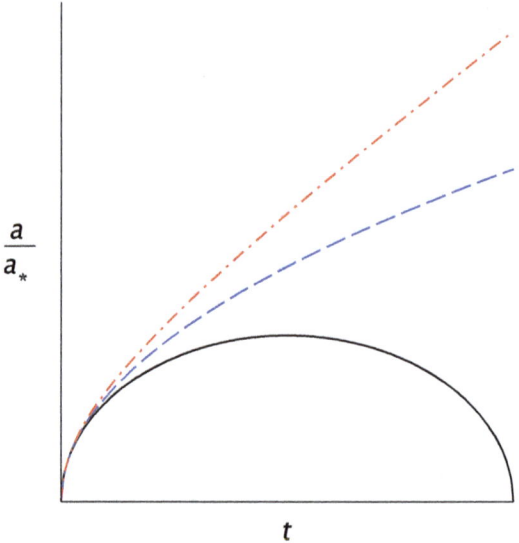

FIG. 1.3. The scale factors for radiation–dominated FLRW universes: positively curved, spatially flat, and negatively curved.

Radiation–dominated closed universes corresponding to $\kappa = +1$ exist only for a finite time. They expand from a Big Bang to a maximum size and then collapse, ending in a "Big Crunch" singularity (this is also a true spacetime singularity where curvature invariants diverge together with the energy density). The contracting phase is the time–reversal of the expanding phase. Open universes, corresponding to $\kappa = -1$, begin at a Big Bang and expand forever. The spatially flat universe ($\kappa = 0$) marks the threshold between these two behaviours. It also expands forever from a Big Bang, but the scale factor arrives to $t = +\infty$ with zero slope since $\dot{a} \simeq 1/t \to 0$ as $t \to +\infty$.

Approaching the Big Bang as $t \to 0^+$, the exact solutions $a(t)$ obtained for $\kappa = \pm 1$ and $\Lambda = 0$ go over to the corresponding $\kappa = 0$ solutions with the same equation of state. This is true not only for radiation solutions, but also for other constant barotropic equations of state if $w > -1/3$. However, the early universe was hot and dominated by photons and ultra–relativistic particles and the use of the radiation equation of state in the early universe is appropriate.

The prediction of the Big Bang in FLRW universes is more general than the specific solutions examined, and it requires only to postulate energy conditions. Restricting ourselves to a perfect fluid[13] and assuming the strong energy condition $\rho + 3P > 0$, note that the scale factor is positive, then the acceleration equation

$$\frac{\ddot{a}}{a} = -\frac{4\pi G}{3}(\rho + 3P) \tag{1.223}$$

implies that the universe is decelerated, $\ddot{a} < 0$ when $w > -1/3$. Since the universe expands, it is $\dot{a} > 0$.

Since $a > 0, \dot{a} > 0$ and $\ddot{a} < 0$, the function $a(t)$ increases and has concavity facing downward and, since it is continuous, it must intersect the t–axis at a finite time t_*. This is the prediction of the Big Bang. The argument holds for all values of the curvature index and does not depend on the equation of state, as long as $\rho + 3P > 0$.

Furthermore, both the slope of the scale factor and the energy density

$$\rho(a) = \frac{\rho_*}{a^{3(w+1)}} \tag{1.224}$$

diverge at the Big Bang. For $w > -1/3$, $\rho(a)$ diverges faster than the curvature term $-\kappa/a^2$ in the Friedmann equation (1.179), that gives

$$\dot{a}^2 = \frac{8\pi G\rho_*}{3}\frac{1}{a^{3w+1}} - \kappa \approx \frac{8\pi G\rho_*}{3\,a^{3w+1}}; \tag{1.225}$$

with $3w + 1 > 0$, $\dot{a}$ diverges at the Big Bang.

This Big Bang prediction fails if the strong energy condition is violated, as is the case of a positive cosmological constant effective fluid with $\rho_\Lambda + 3P_\Lambda = -\Lambda/(4\pi G) < 0$. The spatially flat universe corresponding to this equation of state is the exponential de Sitter solution (1.206) with no Big Bang, beginning in the infinite past.

1.4.3 *FLRW Lagrangian and Hamiltonian*

The fact that, in FLRW cosmology, the Einstein equations reduce to ordinary differential equations make a Lagrangian and a Hamiltonian description possible for the Einstein–Friedmann equations. A Lagrangian for FLRW cosmology is

$$L(a, \dot{a}) = 3a\dot{a}^2 + 8\pi G\,a^3\rho - 3Ka, \tag{1.226}$$

where $\rho = \rho(a)$ is specified by the barotropic equation of state $P = P(\rho)$ and the conservation equation (1.180). Since $\rho = \rho(a)$, the Lagrangian (1.226)

---

[13]See Refs. [202, 435] for a general discussion.

does not depend explicitly on the time t and the corresponding Hamiltonian is conserved,

$$\mathcal{H} = \frac{\partial L}{\partial \dot{a}} \dot{a} - L = 3 a \dot{a}^2 - 8\pi G a^3 \rho + 3Ka = C. \qquad (1.227)$$

These Lagrangian and Hamiltonian solve the inverse variational problem of finding an action integral, however this is not sufficient to complete the derivation. Since the dynamics of general relativity is constrained [82, 435], in the FLRW geometry the Hamiltonian constraint (time–time component of the Einstein equations) imposes that $C = 0$ and is the same as the Friedmann equation (1.179) [435].

1.5 Symmetries of the Einstein–Friedmann equations for spatially flat universes

Symmetries are important in systems with a finite number of degrees of freedom because they are associated with first integrals and offer an avenue for solving their equations of motion. For a spatially flat FLRW universe sourced by a single perfect fluid with constant equation of state $P = w \rho$, the Einstein–Friedmann equations enjoy certain symmetries, which have attracted considerable attention ([3, 28, 32, 88, 88, 95, 96, 111, 135, 135, 151, 151, 179, 314–317, 320, 395, 397, 406, 450] and references therein). In the symmetries of interest in this book, the time t, scale factor a, and Hubble function H are rescaled, changing the cosmic fluid in such a way that the Einstein–Friedmann equations (1.179)–(1.187) are left invariant in form.

The first symmetry [135]

$$a \to \tilde{a} = \frac{1}{a}, \qquad (1.228)$$

$$w \to \tilde{w} = -(w + 2), \qquad (1.229)$$

maps an expanding universe into a contracting one and *vice–versa*.

A different one–parameter group of symmetry transformations of a spatially flat FLRW universe, parametrized by the real number $s \neq 0$, is given by [151]

$$a \to \bar{a} = a^s, \qquad (1.230)$$

$$dt \to d\bar{t} = s \, a^{\frac{3(w+1)(s-1)}{2}} dt, \qquad (1.231)$$

$$\rho \to \bar{\rho} = a^{-3(w+1)(s-1)} \rho. \qquad (1.232)$$

The third type of symmetry of interest here, again for a $\kappa = 0$ universe, is [88, 118]

$$\rho \to \bar{\rho} = \bar{\rho}(\rho) \,, \tag{1.233}$$

$$H \to \bar{H} = \sqrt{\frac{\bar{\rho}}{\rho}} \, H \,, \tag{1.234}$$

$$P \to \bar{P} = -\bar{\rho} + \sqrt{\frac{\rho}{\bar{\rho}}} \, (P + \rho) \, \frac{d\bar{\rho}}{d\rho} \,, \tag{1.235}$$

where the function $\bar{\rho}(\rho)$ has the same sign as ρ and is regular. This symmetry also applies when the barotropic perfect fluid in the FLRW universe does not have linear or constant equation of state.

When a certain physical system is analogous to a spatially flat FLRW universe, one of the cosmological symmetries above may reveal a previously unknown symmetry, and a mathematical structure, of the equations ruling the analogous system. We will apply these symmetries to the equation describing the freezing of bodies of water in winter in Chap. 3, and to the Vialov equation ruling the shapes of glaciers in Chap. 6.

1.6   The accelerating universe and dark energy

In 1917 Einstein introduced the cosmological constant term in the field equations of general relativity. His motivation for doing this was cosmological: relativistic cosmology was just being developed and, at that time, Einstein believed the universe to be static (the Hubble discovery of the redshift of galaxies was not yet known widely). The Einstein–Friedmann equations (1.179)–(1.187) with dust or radiation and $\Lambda = 0$ do not admit static solutions. This realization led Einstein to modify his field equations by adding the only possible term that is compatible with general covariance and with the requirement that the equations be at most of second order in the metric tensor, that is, a constant multiplying the metric tensor [123].

If $\Lambda > 0$, the introduction of the cosmological term has the effect that a static solution of the Einstein–Friedmann equations with dust and positive curvature can be found, which is the Einstein static universe. It is a closed universe with line element

$$ds^2 = -dt^2 + d\chi^2 + \sin^2 \chi \, d\Omega^2_{(2)} \,. \tag{1.236}$$

When, later on, Hubble showed that the spectra of galaxies are, on average, redshifted, it became difficult to retain the cosmological term in the Einstein equations. In addition, Arthur Eddington showed that the Einstein static universe is unstable against perturbations [122], a fatal flaw for a realistic model of the cosmos. This instability is hardly surprising if one considers that the Einstein static universe is achieved by a delicate balance between the attraction of dust and the repulsion of Λ. If the size of the Einstein static universe is increased, the attraction of the now diluted dust decreases, while the repulsion of the positive cosmological constant increases. *Vice–versa*, if the size of the Einstein universe is decreased, the attraction of dust increases and the repulsion of Λ decreases. In both cases, the perturbed universe runs away from its original Einstein static state.

What is more, already in 1917 the Dutch cosmologist Wilhelm de Sitter had found the $\kappa = 0$ solution (1.206) now called "de Sitter spacetime" which is empty and, due to the positive cosmological constant, expands exponentially fast. Therefore, the cosmological constant does not always make the universe static, as hoped by Einstein.

In modern cosmology the de Sitter solution has become even more important because in many models of the early universe and of the late universe it plays the role of an attractor in the phase space of solutions of the Einstein–Friedmann equations [5].

By 1931, the expansion of the universe was established by the work of Hubble and his collaborators and Einstein abandoned the cosmological constant. In the 1960s the Russian astrophysicist Yakov B. Zel'dovich, following ideas from the Russian theoretical physics school expressed by Erast Gliner, realized that quantum vacuum behaves as a cosmological constant. It is, therefore, regarded as a form of energy, the energy of quantum vacuum, and not as a geometric term. This return of the cosmological constant into physics is accompanied by the very serious cosmological constant problem. When Λ is estimated from basic quantum mechanical principles, the numerical value of its cosmological energy density turns out to be about 10^{120} times the critical density, which is instead the energy density measured in our universe. This huge number is derived by the following naive back–of–the–envelope calculation.

Consider the relation between the energy E and the three–momentum $\vec{p}$ of the relativistic free particle (which is nothing but the normalization $p_\mu p^\mu = -m^2$ of the particle's four–momentum p^μ)

$$E^2 = p^2 + m^2 \,. \tag{1.237}$$

A quantum field in Minkowski spacetime, for example a scalar field, is a collection of modes, each one of which is described by a three–dimensional wave vector $\vec{k}$ and has energy $E_k = \sqrt{k^2 + m^2}$. By integrating over all modes in $\vec{k}$–space, the total energy density is

$$\rho_\Lambda = \iiint_{\mathbb{R}^3} d^3\vec{k} \sqrt{k^2 + m^2}$$

$$= \int_0^{2\pi} d\varphi \int_0^\pi d\theta \int_0^{+\infty} dk \, k^2 \sin\theta \sqrt{k^2 + m^4} \,, \qquad (1.238)$$

where we passed to polar coordinates in $\vec{k}$–space and where we can neglect $m^2 \ll k^2$ for large k. This integral diverges as k^4. To cure this divergence, one introduces a cutoff and stops the integration at a maximum value of the momentum k corresponding to a wavelength of the order of the Planck length $\ell_{pl} \simeq 10^{-33}$ cm. This Planck scale cutoff eliminates the ultraviolet divergence but still produces the above–mentioned energy density $\rho_\Lambda = 10^{120}\rho_c$. Even introducing a cutoff at the lower energy scale of grand unified theories, one obtains $\rho_\Lambda \simeq 10^{40}\rho_c$, still a far cry from the observed value.

These calculations are in violent conflict with the cosmological observations providing $\rho_0 \simeq \rho_c$ for our universe. The cosmological constant problem has been termed the most serious problem of modern theoretical physics.

In 1998 it was discovered that the expansion of the universe is accelerated, which can be explained within the context of the Einstein–Friedmann equations by postulating the existence of a mysterious form of dark energy accounting for approximately 70% of the energy content of the universe and with negative pressure and an effective equation of state $P = w\rho$, with w close to -1. The matter content of the universe, including baryons and dark matter, amounts to only approximately 30% of the critical density. On another front, the analysis of the Doppler peaks[14] in the cosmic microwave background and large scale structure surveys indicate that our universe has an energy density equal to the critical density. There are also theoretical arguments why the universe should have critical density. First, it is widely accepted (although not proven) that the early universe underwent a very brief epoch of inflation, *i.e.*, superluminal accelerated expansion. A robust prediction of inflation is that the universe is taken by this expansion to a state extremely close to a spatially flat universe with critical density. Second, the theory of the formation of structures in the universe requires non–relativistic ("cold") dark matter and the presence of a cosmological

---

[14]A Fourier expansion of the temperature anisotropies $\delta T/T (\vartheta, \varphi)$ in the microwave sky.

constant or similar form of energy slowly evolving or constant in time, uniform in space, that does not cluster strongly as baryons do. This so–called ΛCDM model (for Λ–Cold Dark Matter) based on general relativity, dark energy, and dark matter, is currently the standard model of cosmology and incorporates the standard Big Bang cosmology augmented by inflation in the very early history of the universe and by the dominance of a cosmological constant or of dark energy in the late universe, to cause the present acceleration.

The cosmological constant is an obvious candidate for dark energy, but it carries with it the cosmological constant problem. The prevailing attitude in the cosmological community is that it is very unlikely that the cosmological constant is cancelled to one part in 10^{120}, while leaving behind an extremely small residual just sufficient to explain the acceleration of the universe today. Theorists prefer to believe that a yet to be discovered mechanism cancels the cosmological constant exactly, or shelters the universe from its gravity, and that the dark energy is a completely different form of energy. Popular models of dark energy (see [5] for a review) are based on dynamical scalar fields or on modifications of Einstein gravity, but many other models have been proposed since 1998. An observational test to discriminate between the cosmological constant and dynamical dark energy would be the detection of the evolution of the equation of state parameter with redshift, $w = w(z)$, which would rule out the cosmological constant as a candidate for dark energy.

Chapter 2

Cosmic analogies in mathematics

> *Mathematics is the study of analogies between analogies. All science is. Scientists want to show that things that don't look alike are really the same. That is one of their innermost Freudian motivations. In fact, that is what we mean by understanding.*
>
> Giancarlo Rota

2.1 Inverse variational problems

Cosmic analogies can be useful in the solution of the inverse variational problem (a rather general problem of mathematics and physics) for the analogous systems.

The direct variational problem of variational calculus consists of finding the Euler–Lagrange equations and solving them, given an action integral

$$S = \int dt\, L\left(t, x^i, \dot{x}^i\right), \qquad (2.1)$$

where $L\left(t, x^i, \dot{x}^i\right)$ $(i = 1, 2, ..., n)$ is the Lagrangian.

The inverse variational problem (or Helmoltz problem) is the following: given a set of second order ordinary differential equations, find a Lagrangian such that its Euler–Lagrange equations reproduce the given system. There are necessary and sufficient conditions ("Helmoltz conditions") for the inverse variational problem to admit a solution. However, in general, determining whether these conditions are satisfied may imply solving a large system of equations.

To be more precise, consider the system of second order ordinary differential equations

$$\ddot{x}^i = F^i\left(t, x^j(t), \dot{x}^j(t)\right), \qquad i, j = 1, 2, \, ..., \, n, \qquad (2.2)$$

where an overdot denotes differentiation with respect to the independent variable t. The question arises of whether the equations (2.2) are solutions of the Euler–Lagrange equations

$$\frac{d}{dt}\left(\frac{\partial L}{\partial \dot{x}^i}\right) - \frac{\partial L}{\partial x^i} = 0 \qquad (2.3)$$

for some Lagrangian $L\left(t, x^i, \dot{x}^i\right)$. Since, in general, the right hand sides of Eqs. (2.2) depend on the velocities $\dot{x}^i$ as well as on the positions x^i (and, possibly, on t) the Euler–Lagrange equations expand to the form

$$\frac{\partial^2 L}{\partial t \, \partial \dot{x}^i} + \frac{\partial^2 L}{\partial x^j \, \partial \dot{x}^i}\dot{x}^j + \frac{\partial^2 L}{\partial \dot{x}^i \, \partial \dot{x}^j}\ddot{x}^j = 0. \qquad (2.4)$$

This is not the usual form of the Euler–Lagrange equation encountered in most textbooks. The inverse variational problem is reduced to finding a "multiplier" invertible matrix $g_{ij}\left(t, x^\ell, \dot{x}^\ell\right)$ such that

$$g_{ij}\left(\ddot{x}^j - F^j\right) = \frac{d}{dt}\left(\frac{\partial L}{\partial \dot{x}^i}\right) - \frac{\partial L}{\partial x^i} \qquad (2.5)$$

(Sonin–Douglas equations). The matrix g_{ij} must be invertible on some domain where these equations are satisfied. The necessary and sufficient conditions for this to happen (Helmoltz conditions) [113, 209, 359] are rather involved. Following the summary of Ref. [337], we first define the quantities

$$\Gamma^i_j \equiv -\frac{1}{2}\frac{\partial F^i}{\partial x^j}, \qquad (2.6)$$

$$\hat{\Gamma} \equiv \frac{\partial}{\partial t} + \dot{x}^i\frac{\partial}{\partial x^i} + F^i\frac{\partial}{\partial \dot{x}^i}, \qquad (2.7)$$

$$\Phi^i_j \equiv -\frac{\partial F^i}{\partial x^j} - \Gamma^k_j\Gamma^i_k - \hat{\Gamma}\left(\Gamma^i_j\right); \qquad (2.8)$$

then the Helmoltz conditions are [113, 209, 337, 359]

$$g_{ji} = g_{ij} , \qquad (2.9)$$

$$\hat{\Gamma}\left(g_{ij}\right) = g_{ik}\Gamma^k_j + g_{jk}\Gamma^k_i , \qquad (2.10)$$

$$g_{ik}\Phi^k_j = g_{jk}\Phi^k_i , \qquad (2.11)$$

$$\frac{\partial g_{ij}}{\partial \dot{x}^k} = \frac{\partial g_{ik}}{\partial \dot{x}^j} . \qquad (2.12)$$

If a solution g_{ij} exists, then the sought–for Lagrangian is given by

$$\frac{\partial^2 L}{\partial \dot{x}^i \, \partial \dot{x}^j} = g_{ij} . \qquad (2.13)$$

Of course, the inverse variational problem does not always admit a solution (see, *e.g.*, Ref. [337]) and, when it does, the procedure to find the solution explicitly can be quite difficult (*e.g.*, [451]). However, for lower–dimensional problems, the solution can become easier.

Analogies offer an alternative approach to the Helmoltz problem that is, however, only applicable to systems analogous to another system with a known Lagrangian. In our case, these are the systems analogous to FLRW cosmology, which are the subject of this book. Since we know the Lagrangian (and the Hamiltonian) for a FLRW universe, we just "copy" this Lagrangian and write it in terms of the analogous variables on the other side of the analogy. Since the Einstein–Friedmann equations only describe the dynamics of the scale factor $a(t)$, while the energy densities of the perfect fluids filling the universe are determined in terms of a by their conservation equations, we have effectively a one–dimensional problem and systems analogous to FLRW universes can only be one-dimensional. Even with the limitation that the corresponding system must reduce to a single ordinary differential equation of second order, the cosmic analogy offers an unusual method to solve the inverse variational problem for the analogous systems.

2.2   When the same equation describes many phenomena

Here we discuss an ordinary differential equation that contains the Friedmann equation as a special case and turns out to be useful in our quest for cosmic analogies.

There are many physical and natural phenomena that can be described, at least approximately, by the first order ordinary differential equation

$$\dot{y}(t) = f(y(t)),\qquad(2.14)$$

where an overdot denotes differentiation with respect to the independent variable. Since in many realizations the latter describes time, we will denote it with the letter t. The functions $y(t)$ and $f(y)$ are assumed to be as regular as needed in the physical problems studied. Physical phenomena described by Eq. (2.14) include the one-dimensional motion of a particle under a conservative force, for which the square of Eq. (2.14) is the energy integral of relativistic cosmology, in which the Friedmann equation has the form of the square of (2.14); and one–dimensional systems in thermal physics, fluid mechanics, and relativistic gravity [388]. Natural phenomena described by this equation are encountered in seismology, oceanography, glaciology, and other areas [388]. Some of them are discussed in the following chapters.

It is convenient to take a look at the peculiar mathematical properties of Eq. (2.14), which in turn dictate interesting physical properties of the physical systems described by it. First, this equation is separable and is formally integrated to

$$I \equiv \int \frac{dy}{f(y)} = t - t_0,\qquad(2.15)$$

where t_0 is an integration constant. It is not always possible to express this integral explicitly in terms of elementary functions and often, even when this happens, it may not be possible to invert explicitly Eq. (2.15) to obtain $y(t)$ (in many situations this is immaterial). Since many natural phenomena in very different contexts all obey the formal equation (2.14) [20, 41, 84, 121, 141, 142, 147, 154, 159, 160, 166, 305, 343, 345, 360], many formal analogies between these very different physical systems are possible, although only a few have ben discussed in the literature.

The inverse variational problem of obtaining a variational principle and a Lagrangian function that reproduce Eq. (2.14) has a solution. In spite of the fact that Eq. (2.14) is of first order, it can be given Lagrangian and Hamiltonian formulations—at a price. When this equation is promoted from first to second order in this process, one introduces an extra solution not contained in the original first order equation. The dimension of the phase space and the number of initial conditions are doubled rather arbitrarily and extra information must be added to eliminate the spurious solution.

Let us begin by considering the following special case, which is of interest in building cosmological analogies in the following chapters. If the function $f(y)$ is quadratic in y, Eq. (2.14) assumes the form

$$\dot{y}(t) = a_0 + b_0\, y(t) + c_0\, y^2(t)\,, \qquad (2.16)$$

where a_0, b_0, and c_0 are constant coefficients. This is a special case of the Riccati equation of general form [210, 222]

$$\dot{y}(t) = a(t) + b(t)\, y(t) + c(t)\, y^2(t) \qquad (2.17)$$

with time–dependent coefficients. If $a_0 = 0$, Eq. (2.16) becomes homogeneous and reduces to a special case of the Bernoulli equation, the general form of which is Eq. (2.17) with $a(t) = 0$ [42, 61, 210, 222].

Another situation relevant for FLRW cosmology occurs when the first order equation takes the form

$$\dot{y}(t) = \sqrt{h(y(t))} \qquad (2.18)$$

and in this case it may be convenient to square the equation, which is often analogous to the Friedmann equation (1.179) for some specific FLRW universe.

Let us examine the zeros of $f(y)$, which are constant solutions of the equation $\dot{y} = f(y)$, and their stability [144].

First, note that if $f(y) \neq 0$ for all values of y, no constant solution of Eq. (2.14) can exist. Then, consider the possibility that $f(y)$ has a zero at y_c: then the equation admits the constant solution

$$y(t) = y_c \qquad \forall t\,, \qquad (2.19)$$

which is an equilibrium point of this simple dynamical system. Conversely, all the constant solutions $y_c = $ const. correspond to zeros of $f(y)$.

2.2.1 *Stability of the equilibrium solutions $y = y_c$*

In order to study the stability of these constant solutions or equilibrium points when $f(y)$ has zeros, consider first the simplest situation in which $f(y)$ has a single zero.

Single isolated zero of $f(y)$. Assume that $f(y_c) = 0$ and $f'(y_c) \neq 0$ (where $f'(y) \equiv df/dy$), in which case only two situations can occur:

- When $f'(y_c) > 0$, by continuity we have $\dot{y} = f(y) < 0$ for $y < y_c$ and $\dot{y} > 0$ for $y > y_c$ in a neighbourhood of y_c. The phase space (t, y) is one–dimensional and, if $y < y_c$, a nearby solution $y(t)$ decreases, getting further and further away from the equilibrium solution $y(t) = y_c$. If $y > y_c$, instead, nearby solutions starting out with $y > y_c$ increase and depart from the $y = y_c$ equilibrium. Therefore, the constant solution $y(t) = y_c$ is unstable and repels nearby trajectories in phase space.

• The second possibility occurs if $f'(y_c) < 0$, in which case the situation is reversed. A solution $y(t)$ starting out below the equilibrium solution y_c and nearby increases and approaches it, while nearby solutions starting above with $y > y_c$, decrease and approach equilibrium. In this case, the equilibrium solution $y(t) = y_c$ is stable and attracts nearby orbits in phase space.

Double root of $f(y) = 0$. Let us discuss now the situation in which y_c is a double root of the equation $f(y) = 0$. In other words, we now have $f(y_c) = f'(y_c) = 0$ and $f''(y_c) \neq 0$ is either positive or negative.

• If $f''(y_c)$ is positive, $f(y)$ has a maximum at y_c and, according to our differential equation, it is $\dot{y} = f(y) < 0$ in a neighbourhood of y_c. Phase space trajectories starting with $y < y_c$ near the equilibrium solution $y \equiv y_c$ must decrease and depart from it, while they become closer if they start above it at some initial $y > y_c$. In this case the equilibrium solution $y \equiv y_c$ is semi–stable.

• Finally, when $f''(y_c) > 0$, the function $f(y)$ is minimum at y_c and $\dot{y} = f(y) > 0$ in a neighborhood of this point, hence the conclusions of the previous situation are reversed, the solution $y = y_c$ still being semi–stable.

In many situations of physical interest, only one of the scenarios $y > y_c$ or $y < y_c$ is physically interesting and the problem is streamlined.[1]

Higher order roots of $f(y) = 0$. It is in principle possible that $f(y)$, $f'(y)$ and higher order derivatives all vanish at some point y_c. In this case, the lowest order derivative $f^{(n)}(y)$ that does not vanish at y_c must be examined. If n is odd, then $f^{(n)}(y_c) < 0$ implies stability and $f^{(n)}(y_c) > 0$ implies instability. If, instead, n is even, there is stability on one side and instability on the other side of the equilibrium solution $y(t) = y_c$.

---

[1] This situation occurs, for example, in the densification of glacier firn where ice grains grow in size, which is reported in Chap. 6.

2.2.2 Symmetry for power–law $f(y)$

In cosmology and in many physical systems for which an analogy between a first order differential equation and the Friedmann equation holds, the function $f(y)$ on the right hand side of Eq. (2.14) is a power–law, producing

$$\dot{y} = f_0\, y^\alpha\,, \tag{2.20}$$

where f_0 and α are constants. Then, as shown in Chap. 1, Eq. (2.20) is invariant under the symmetry operation

$$y \longrightarrow \bar{y} = y^s\,, \tag{2.21}$$

$$dt \longrightarrow d\bar{t} = s\, y^{(s-1)(1-\alpha)} dt\,, \tag{2.22}$$

where $s \neq 0$. In fact, we have

$$\frac{d\bar{y}}{d\bar{t}} = \frac{s\, y^{s-1} dy}{s\, y^{(s-1)(1-\alpha)} dt} = y^{(s-1)\alpha}\, \frac{dy}{dt} = y^{(s-1)\alpha}\, f_0\, y^\alpha = f_0\, \bar{y}^\alpha\,. \tag{2.23}$$

The operations (2.21) and (2.22) form a one–parameter group of symmetries with respect to the composition of maps labelled by the parameter s because:

- Two operations (2.21), (2.22) give another operation of the same type since

$$y \to \bar{y} = y^s \to \bar{\bar{y}} = \bar{y}^p = y^q\,, \tag{2.24}$$

and

$$dt \to d\bar{t} = s\, y^{(s-1)(1-\alpha)} dt \to d\bar{\bar{t}} = p\, \bar{y}^{(p-1)(1-\alpha)} d\bar{t}$$

$$= q\, y^{(1-\alpha)(q-1)} dt \tag{2.25}$$

with $q = p\,s$.
- The identity corresponding to $s = 1$ is the neutral element of this group.
- The inverse of the transformation identified by the parameter $s \neq 0$ is another symmetry of the kind (2.21), (2.22) with parameter $1/s$.
- The operation is associative.

The group is commutative: the final result of the composition of two consecutive transformations does not depend on the order in which these operations are applied since $sp = ps = q$ in Eqs. (2.24) and (2.25).

2.2.3   *Lagrangian and Hamiltonian formulations*

The Lagrangian and Hamiltonian formulations of the equation $\dot{y} = f(y)$ are easily obtained. Its differentiation yields $\ddot{y} = f'(y)\,\dot{y}$ and then (2.14) gives

$$\ddot{y} - f'(y)f(y) = 0\,. \tag{2.26}$$

By differentiating the original equation we have raised the order by one, which means that the general solution of Eq. (2.26) now contains two independent modes, and two integration constants, instead of the single one of our original problem, corresponding to initial conditions $\left(y_0, \dot{y}_0\right)$ instead of y_0. Keeping this fact in mind, one notes that

$$L\left(y, \dot{y}\right) = \dot{y}^2 + f^2(y) \tag{2.27}$$

is a Lagrangian for Eq. (2.26). Indeed, since

$$\pi_y \equiv \frac{\partial L}{\partial \dot{y}} = 2\,\dot{y}\,, \tag{2.28}$$

$$\frac{\partial L}{\partial y} = 2\,f(y)f'(y)\,, \tag{2.29}$$

where π_y denotes the momentum canonically conjugated to the variable y, it is straightforward to see that the Euler–Lagrange equation

$$\frac{d}{dt}\left(\frac{\partial L}{\partial \dot{y}}\right) - \frac{\partial L}{\partial y} = 0 \tag{2.30}$$

gives back Eq. (2.26).

The Hamiltonian

$$\mathcal{H} = \pi_y\,\dot{y} - L = \dot{y}^2 - f^2(y) \tag{2.31}$$

obtained from the Lagrangian (2.27) is conserved because it does not depend explicitly on the time t [183], which gives the first integral of motion

$$\dot{y}^2 - f^2(y) = \text{const.} \tag{2.32}$$

This conservation equation does not coincide with the original Eq. (2.14), which can only be recovered by providing the extra information

$$\mathcal{H} = 0\,. \tag{2.33}$$

This extra input can be imposed as a first order constraint on the dynamics and the need to provide it is due to the fact that, passing from first to second order, the space of solutions of the original problem has been enlarged, introducing spurious solutions. Taken at face value, the Lagrangian (2.27) or the Hamiltonian (2.31), produces a genuine second order ordinary differential equation.

We now see that, in addition to the equation (2.14) that we started with, the new conservation equation (2.33) contains also its spurious sister

$$\dot{y} = -f(y), \tag{2.34}$$

and the Lagrangian (2.27) or the Hamiltonian (2.31) do not distinguish f and $-f$. However, inverting the sign in the right hand side of the original equation (2.14) makes a world of difference for its solutions from both the mathematical and the physical points of view. In the situations in which the equations $\dot{y} = \pm f(y)$ describe a physical process and its time–reverse, the solution (for example, the motion of a particle) breaks the symmetry: the initial condition selects one or the other of the two equations $\dot{y} = \pm f(y)$. The first integral $\mathcal{H} = 0$ only says that the total energy vanishes (it assigns an initial condition on the energy and fixes its value forever) but does not resolve the ambiguity in choosing one of the two equations $\dot{y} = \pm f(y)$, while the initial condition $\dot{y}_0$ does.

2.2.4 *Mechanical analogy*

There is a rather obvious mechanical analogy between Eqs. (2.31) and (2.33) and the one–dimensional motion of a point particle in a conservative force. We assign to this fictitious particle unit mass, position $y(t)$ at time t, kinetic energy $\dot{y}^2(t)/2$, and a conservative force $F(y) = -dV/dy$ with potential energy $V(y) = -f^2(y)/2$. This potential energy is always negative and unbounded from below, corresponding to an inverted harmonic oscillator.

The total mechanical energy of the fictitious particle vanishes,

$$E \equiv \frac{\dot{y}^2}{2} + V(y) = 0. \tag{2.35}$$

Graphically, this constraint corresponds to the intersection between the horizontal line $V = E$ and the graph of the potential energy $V(y)$ precisely at $E = 0$. This intersection determines the possible turning points of the motion (if they exist), where the motion of the particle stops and its kinetic energy vanishes. Since it is always $V(y) \leq 0$, the only turning points correspond to the zeros of $V(y)$ and of $f(y)$. These turning points are the only possible positions of equilibrium of the particle and the only constant solutions of Eq. (2.14).

One may, erroneusly, be led to conclude that this equilibrium solution $y \equiv y_c$ is always unstable because $V(y)$ has a maximum there, but this conclusion would contradict our previous discussion and is incorrect. Indeed, the potential $V(y)$ obtained by squaring Eq. (2.14) describes both equations $\dot{y} = \pm f(y)$, and the essential information on the sign of $f(y)$ and of $\dot{y}$ is not contained in V. When the sign of the right hand side of Eq. (2.14) is inverted, growth is turned into decay and *vice–versa*. The stability of the solutions of this equation is changed drastically if $f(y) < 0$. We conclude that the point–particle analogy for the equation $\dot{y} = f(y)$ is only appropriate if $f(y) \geq 0$.

2.3   Exponential growth and decay

Many physical phenomena are described by exponential growth and decay, which correspond to the very simple linear form of Eq. (2.14). Examples are the decay of radioactive isotopes, of radioactive natural rocks (*e.g.*, [330]), the attenuation of radiation propagating through a medium [47] (including light penetrating the atmosphere [334, 428]), the multiplication of biological cells, the increasing mass of algae decorating the surface of a pond, the spread of a forest fire, and many others. In these situations, $f(y)$ in Eq. (2.20) is linear and

$$\dot{y} = \pm k\, y \,, \tag{2.36}$$

where $k > 0$ is constant and $\alpha = 1$ (we have already seen that the trivial case $k = 0$ produces only constant solutions). The Lagrangian (2.27) and the Hamiltonian (2.31) of the inverse variational problem read

$$L = \dot{y}^2 + k^2 y^2 \tag{2.37}$$

and

$$\mathcal{H} = \dot{y}^2 - k^2 y^2 \,. \tag{2.38}$$

The constraint $\mathcal{H} = 0$ allows both growth and decay, not distinguishing between these possibilities. Were this constraint not valid, one would have instead the equation of motion $\mathcal{H} = C$, with C positive, negative, or zero and one would obtain the first integral $\dot{y} = \pm\sqrt{C + k^2 y^2}$ which is substantially more general than the special situation for $C = 0$ giving (2.36). Only asymptotically at late times would the solution of $\dot{y} = \sqrt{C + k^2 y^2}$ approach that of the "true" equation $\dot{y} = k\, y$.

For exponential growth and decay, the symmetry (2.21), (2.22) becomes

$$y \rightarrow \bar{y} = y^s \,, \tag{2.39}$$

$$t \rightarrow \bar{t} = st \,, \tag{2.40}$$

and reduces to the scaling property of the exponentials $y_{\pm}(t) = y_0 \, e^{\pm kt}$:

$$\left(y_0 e^{\pm kt}\right)^s = y_0^s \, e^{\pm skt} = \bar{y}_0 \, e^{\pm k\bar{t}} \,, \tag{2.41}$$

where $y_0 = y(0)$ is the initial value and $d\bar{y}/d\bar{t} = \pm k\,\bar{y}$. The corresponding symmetries of the solutions are manifest in the relations

$$y(t) = y_0 \left(t - t_0\right)^{\frac{1}{1-\alpha}} \,, \qquad y_0 = \left[f_0(1-\alpha)\right]^{\frac{1}{1-\alpha}} \quad \text{if} \quad \alpha \neq 1, \tag{2.42}$$

$$y(t) = y_0 \, e^{f_0 t} \,, \qquad y_0 = e^{-f_0 t_0} \quad \text{if} \quad \alpha = 1. \tag{2.43}$$

In the mechanical analogy discussed above, the potential energy $V(y) = -k^2 y^2 / 2$ of the associated inverted harmonic oscillator is even (*i.e.*, $V(-y) = V(y)$) and is insensitive to the sign choice in $\pm k$. There is a unique equilibrium $y(t) \equiv 0$ consistent with the discussion about the roots of $f(y) = 0$.

The stability of this equilibrium point deserves some discussion. For the equation $\dot{y} = k\,y$ ruling exponential growth, a particle close to $y = 0$ starting at positive initial position y_0 will fall to the right, with its position $y(t)$ growing exponentially in time—this particle runs away from $y = 0$. A different particle beginning at $y_0 < 0$ will instead fall to the left and its position will be $y(t) = -|y_0| \, e^{kt}$, *i.e.*, this particle escapes to minus infinity. For this situation, the mechanical analogy with a point particle based on the equation $\mathcal{H} = 0$ is reliable.

The situation is different for the other equation $\dot{y} = -k\,y$ describing exponential decay. We see that a particle placed initially near $y = 0$ at some $y_0 > 0$ has position $y(t)$ decreasing exponentially fast. This particle approaches $y = 0$ instead of falling to the right. Moreover, any solution starting near zero with $y_0 < 0$ approaches the origin exponentially fast. Therefore, the equilibrium point $y(t) = 0$ is stable. This example demonstrates that the mechanical analogy provides a reliable assessment of stability only when $f(y) \geq 0$ in Eq. (2.14).

Exponential growth described by $\dot{y} = k\,y$ has an almost trivial analogy with FLRW cosmology. Dividing this equation by y and squaring gives

$$\left(\frac{\dot{y}}{y}\right)^2 = k^2 \,. \tag{2.44}$$

Under the correspondence between $y(t)$ and the scale factor $a(t)$, this equation is analogous to the Friedmann equation (1.179) for an analogous universe that is empty but propelled by the cosmological constant $\Lambda = 3k^2$.

The solution is the de Sitter universe with scale factor

$$a(t) = a_0 \, e^{kt} \,, \tag{2.45}$$

where a_0 is a positive constant.

2.4 de Sitter universe and Fibonacci's rabbits

We now begin examining analogies between FLRW cosmology and specific, well–known systems in mathematics, physics, and the earth sciences. Our first analogy involves the famous Fibonacci numbers appearing in many areas of science and in many natural systems, including plants (*e.g.*, in the geometric arrangement of leaves on the stem, in sunflowers, artichokes, pineapples, pinecones), snails, animal populations, the reproduction of rabbits and benthic polyps, and in shells (Fig. 2.1). The golden mean related to Fibonacci numbers has been taken as the most eye-pleasing aspect ratio and has been widely used in art, for example in the Parthenon of Athens, Leonardo da Vinci's Monna Lisa, Le Corbusier's architecture, *etc.*).

Leonardo Pisano, also called Leonardo Bonacci and better known as Fibonacci is remembered for his numerous contributions to mathematics but perhaps his most impactful one was the introduction of Hindu–Arabic numerals to Europe through the *Liber Abaci* published in 1202 [371]. Among his mathematical works, Fibonacci posed, and solved, the problem of determining the population of rabbits and its growth in time under idealized situations. The solution of this problem is now known as the Fibonacci sequence $0, 1, 1, 2, 3, 5, 8, 13, 21, 34, 55, \ldots$ The Fibonacci sequence is given by the recursion relation

$$F_n = F_{n-1} + F_{n-2} \,, \tag{2.46}$$

where $F_0 = 0$, $F_1 = 1$, and F_n is the n-th Fibonacci number. An interesting property of this sequence is that the ratio of two consecutive Fibonacci numbers F_{n+1}/F_n approaches the golden ratio

$$\varphi \equiv \frac{1 + \sqrt{5}}{2} \approx 1.61803398\ldots \tag{2.47}$$

as $n \to +\infty$. The Fibonacci numbers have many applications in mathematics and in the natural sciences, see, *e.g.*, Refs. [2, 23, 50, 51, 176, 230, 267], and various generalizations.

The Binet formula

$$F_n = \frac{\varphi^n - (-\varphi)^{-n}}{\sqrt{5}} \tag{2.48}$$

extends the Fibonacci sequence to the continuum. Another property is that the entire Fibonacci sequence can be reproduced by two analytic functions: the function

$$F_{(e)}(x) = \frac{\varphi^x - \varphi^{-x}}{\sqrt{5}} = \frac{2}{\sqrt{5}} \sinh{(x \ln \varphi)} \qquad (2.49)$$

reproduces half of the Fibonacci numbers according to

$$F_n = F_{(e)}(n) \qquad (2.50)$$

for even $x = n \in \mathbb{N}$, while the other function

$$F_{(o)}(x) = \frac{\varphi^x + \varphi^{-x}}{\sqrt{5}} = \frac{2}{\sqrt{5}} \cosh{(x \ln \varphi)} \qquad (2.51)$$

gives the remaining Fibonacci numbers for odd $x = n \in \mathbb{N}$. Finally, the third function

$$F(x) = \frac{\varphi^x - \cos{(\pi x)} \, \varphi^{-x}}{\sqrt{5}} \,, \qquad (2.52)$$

where $x = n \in \mathbb{N}$, reproduces the entire Fibonacci sequence. The functions $F_{(e,o)}(x)$ (but not $F(x)$) generate analogies with FLRW cosmologies [153].

FIG. 2.1. The Fibonacci sequence appears in a *Nautilus* shell. The logarithmic spiral of this shell is formed by drawing circular arcs that connect the opposite sides of squares, whose areas are multiples of a basic given area. These multiples are the numbers of the Fibonacci sequence.

2.4.1   *The cosmic analogy*

To begin with, we note that the functions $F_{(e,o)}$ obey the equations

$$F'_{(e)}(x) = (\ln \varphi)\, F_{(o)}(x)\,, \tag{2.53}$$

$$F'_{(o)}(x) = (\ln \varphi)\, F_{(e)}(x)\,, \tag{2.54}$$

which are dual to each other, and where a prime denotes differentiation with respect to x. The functions $F_{(e)}$ and $F_{(o)}$ are also linearly independent solutions of the second order ordinary differential equation

$$F''_{(e,o)} - \left(\ln^2 \varphi\right) F_{(e,o)} = 0\,. \tag{2.55}$$

To a physicist, this equation describes the motion in one dimension of a particle with position F subject to the inverted harmonic oscillator potential $V(F) = -kF^2/2$, with the constant $k = 2\ln^2 \varphi$. We have, therefore, a simple mechanical analogy. The inverted harmonic oscillator is often used in physics courses as an example of an unstable mechanical system. By examining the Fibonacci sequence, it is apparent that this property of the analogous oscillator is simply the translation of the fact that the Fibonacci numbers F_n increase without bound as $n \to \infty$.

Let us introduce the function

$$f(x) = \varphi^x\,; \tag{2.56}$$

according to the definition of Ref. [195], this is a *Fibonacci function*, i.e.,

$$f(x + 2) = f(x + 1) + f(x) \qquad \forall x \in \mathbb{R}\,. \tag{2.57}$$

It is straightforward to demonstrate that the only power–law function $y(x) = b^x$ which is a Fibonacci function is the one with base equal to the golden ratio, i.e., $b = \varphi$ [195]. The functions $F_{(e,o)}(x)$ are not Fibonacci functions because they contain φ^{-x}, which is not a Fibonacci function.

Let us discuss now the analogy between Fibonacci numbers and FLRW cosmology. Begin with the function $F_{(e)}(x)$, which solves the first order equation [153]

$$F'_{(e)} = \frac{2}{\sqrt{5}}\,\ln \varphi \cosh (x \ln \varphi) = \frac{2}{\sqrt{5}}\,\ln \varphi \sqrt{1 + \frac{5F_{(e)}^2}{4}}\,. \tag{2.58}$$

If we divide this equation by $F_{(e)}$ and square the result, we have

$$\left(\frac{F'_{(e)}}{F_{(e)}}\right)^2 = \ln^2 \varphi + \frac{4\ln^2 \varphi}{5F_{(e)}^2}\,. \tag{2.59}$$

Under the identification $\left(x, F_{(e)}(x)\right) \to \left(t, a(t)\right)$ and

$$\rho = P = 0 \,, \tag{2.60}$$

$$\Lambda = 3\ln^2\varphi \,, \tag{2.61}$$

$$K = -\frac{4\ln^2\varphi}{5} \,, \tag{2.62}$$

Eq. (2.59) is formally the Friedmann equation (1.179). Contrary to FLRW cosmology, however, the quantities ρ, P, Λ, and K are dimensionless. The formal equivalence with the Friedmann equation (1.179) alone does not guarantee the validity of the analogy, which must be completed by Eqs. (1.180) and (1.187). These equations can be easily shown to hold as well using Eq. (2.55). The conservation equation is satisfied trivially by $\rho = P = 0$ and by the effective Λ–fluid with equation of state $P_\Lambda = -\rho_\Lambda = $ constant. Then, the analogous cosmos has scale factor corresponding to the function $F_{(e)}(x)$, possesses hyperbolic three–dimensional time slices, and is empty. Its expansion is driven by the repulsive gravity of the cosmological constant.

When we squared Eq. (2.58) we introduced contracting solutions (and analogous universes) with $F'_{(e)} < 0$. Correspondingly, the Friedmann equation (1.179) yields

$$\dot{a} = \pm\sqrt{\frac{\Lambda a^2}{3} + |K|} \,, \tag{2.63}$$

which is easily integrated to

$$\ln\left[C\left(\sqrt{\Lambda^2 a^2 + 3\Lambda|K|} + \Lambda a\right)\right] = \pm\sqrt{\frac{\Lambda}{3}}\,(t - t_0) \,, \tag{2.64}$$

with C and t_0 are integration constants.[2] This equation then yields the scale factor

$$a(t) = a_0\left[e^{\pm\sqrt{\frac{\Lambda}{3}}(t-t_0)} - 3|K|\Lambda C^2\, e^{\mp\sqrt{\frac{\Lambda}{3}}(t-t_0)}\right] \,, \tag{2.65}$$

where a_0 is a constant. As usual in FLRW cosmology, the dimensionless radial coordinate r can be rescaled to normalize the curvature index K to

---

[2] Here the constant C is introduced to make the argument of the logarithm dimensionless. In fact, in the units used here [82, 202, 435] the scale factor a and the cosmological constant Λ have dimensions of a length and of an inverse length squared, respectively.

± 1. Here instead we rescale this coordinate so that $|K| \Lambda C^2 = 1/3$, giving

$$a(t) = \pm a_0 \sinh\left[\sqrt{\frac{\Lambda}{3}}(t - t_0)\right].\tag{2.66}$$

Since the scale factor is non–negative and only vanishes at cosmological spacetime singularities, the upper sign applies for $t \geq t_0$ and the lower one for $t \leq t_0$. At late times $t \to +\infty$, this line element asymptotes to the de Sitter one with [153]

$$a(t) = a_0\, e^{\sqrt{\frac{\Lambda}{3}}(t - t_0)}.\tag{2.67}$$

We can now turn to the other function $F_{(o)}(x)$ generating the remaining Fibonacci numbers. Repeating the procedure followed for $F_{(e)}(x)$ yields

$$\left(\frac{F'_{(o)}}{F_{(o)}}\right)^2 = \ln^2 \varphi - \frac{4\ln^2 \varphi}{5 F_{(o)}^2},\tag{2.68}$$

which is also the formal Friedmann equation (1.179) if we set

$$\rho = P = 0,\tag{2.69}$$

$$\Lambda = 3\ln^2 \varphi,\tag{2.70}$$

$$K = \frac{4\ln^2 \varphi}{5}.\tag{2.71}$$

This second analogous FLRW universe is also empty and expands because of the cosmological constant, but now the spatial sections are three–spheres and closed surfaces. Squaring our original equation has introduced the possibility that $\dot{a}$ is negative, which now produces the possible scale factors

$$a(t) = a_0\left[e^{\pm\sqrt{\frac{\Lambda}{3}}(t - t_0)} + 3K\Lambda C^2\, e^{\mp\sqrt{\frac{\Lambda}{3}}(t - t_0)}\right].\tag{2.72}$$

Normalizing the constant K reduces $a(t)$ to the hyperbolic cosine

$$a(t) = a_0 \cosh\left[\sqrt{\frac{\Lambda}{3}}(t - t_0)\right],\tag{2.73}$$

which belongs to a bouncing universe that contracts from infinite size, reaches the minimum size a_0, and then expands forever approaching the de Sitter state at late times.

What's left to discuss is now only the Fibonacci function $f(x) = \varphi^x$, which obeys the equation

$$\left(\frac{f'}{f}\right)^2 = \ln^2 \varphi \, ; \tag{2.74}$$

the (now obvious) cosmological analogy gives an empty de Sitter universe which is spatially flat and expands exponentially fast with scale factor $a = a_0 \exp \left[\sqrt{\frac{\Lambda}{3}} (t - t_0)\right]$ because of the positive cosmological constant $\Lambda = 3 \ln^2 \varphi$.

2.4.2 Lagrangian, Hamiltonian, and a new invariant

Although it is unusual to contemplate Lagrangians and Hamiltonian for Fibonacci numbers, the generating functions obey simple first order differential equations that originate an inverse variational problem [153]. The mechanical analogy based on Eq. (2.55) and the cosmological analogy discussed above suggest the Lagrangian for $F_{(e,o)}(x)$

$$L_{(e,o)}\left(F_{(e,o)}(x), F'_{(e,o)}(x)\right) = \frac{1}{2} \left(F'_{(e,o)}\right)^2 + \frac{\ln^2 \varphi}{2} F^2_{(e,o)}, \tag{2.75}$$

which gives rise to the Euler–Lagrange equation

$$\frac{d}{dx} \left(\frac{\partial L_{(e,o)}}{\partial F'_{(e,o)}}\right) - \frac{\partial L_{(e,o)}}{\partial F_{(e,o)}} = 0 \tag{2.76}$$

and reproduces Eq. (2.55). The corresponding Hamiltonian is a function of the canonical variables

$$q_{(e,o)} \equiv F_{(e,o)}, \tag{2.77}$$

$$p_{(e,o)} \equiv \frac{\partial l_{(e,o)}}{\partial F'_{(e,o)}} = F'_{(e,o)}, \tag{2.78}$$

and reads

$$\mathcal{H}_{(e,o)} = p_{(e,o)} F'_{(e,o)} - L_{(e,o)} = \frac{\left(p_{(e,o)}\right)^2}{2} - \frac{\ln^2 \varphi}{2} F^2_{(e,o)}. \tag{2.79}$$

The Hamilton equations

$$q'_{(e,o)} = \frac{\partial \mathcal{H}_{(e,o)}}{\partial p_{(e,o)}} = p_{(e,o)} = F'_{(e,o)} = \ln \varphi \, F_{(o,e)} \tag{2.80}$$

give back Eqs. (2.53), (2.54), and

$$p'_{(e,o)} = -\frac{\partial \mathcal{H}_{(e,o)}}{\partial q_{(e,o)}} = \ln^2 \varphi \, F_{(e,o)} \, . \tag{2.81}$$

The Lagrangian $L_{(e,o)}$ does not depend explicitly on x and, therefore, the associated Hamiltonian $\mathcal{H}_{(e,o)}$ is conserved,

$$\frac{\left(p_{(e,o)}\right)^2}{2} - \frac{\ln^2 \varphi}{2} F^2_{(e,o)} = E_{(e,o)} \, . \tag{2.82}$$

The constants $E_{(e,o)}$ represent the energy of the system. When these are expressed explicitly as

$$E_{(e,o)} = \frac{\ln^2 \varphi}{2} \left(F_{(o,e)} + F_{(e,o)} \right) \left(F_{(o,e)} - F_{(e,o)} \right) = \pm \frac{2 \ln^2 \varphi}{5}, \tag{2.83}$$

we identify a first integral $\left(F^2_{(o,e)} - F^2_{(e,o)} \right)$ on the Fibonacci side of the analogy, which leads us to a corresponding first integral in terms of the discrete Fibonacci sequence [153]. We can then state that[3] [153] *the quantity*

$$I = \left[(F_{2m} + F_{2m-1})^2 - (F_{2m-1} + F_{2m-2})^2 \right]$$

$$= F^2_{2m} + F^2_{2m-1} - F^2_{2m-1} - F^2_{2m-2} + 2 \left(F_{2m} F_{2m-1} - F_{2m-1} F_{2m-2} \right) \tag{2.84}$$

is an invariant of the Fibonacci sequence (i.e., it does not depend on m).

To conclude, the generating functions $F_{(e,o)}$ for the continuum generalization of the Fibonacci sequence obey elementary first order differential equations that lead to a mechanical analogy with an inverted harmonic oscillator and to analogies with FLRW universes. In turn, these lead naturally to an inverse variational problem and to a Lagrangian and a Hamiltonian description for the equations obeyed by $F_{(e,o)}$. The conservation of the Hamiltonian brings to light the invariant I expressed by Eq. (2.84) for the discrete Fibonacci sequence. Although the latter may have limited or no practical interest, the cosmological analogy brings something new to mathematics that originated in the middle ages.

---

[3]The invariant (2.84) is not related to other known invariants (such as those of the Fibonacci convolution sequences [215]).

2.5 Cosmic analogues of the logistic equation

Our next analogy with FLRW cosmology involves the renowned logistic equation, which originated in population dynamics but is nowadays applied in many other areas of science. The logistic equation

$$\dot{f}(t) = r f(t) \left[1 - f(t) \right], \tag{2.85}$$

where the parameter r is a positive constant and an overdot denotes differentiation with respect to the independent variable t, has as a solution the well known sigmoid

$$f(t) = \frac{f_0\, e^{rt}}{1 + f_0\, e^{rt}}, \tag{2.86}$$

where f_0 is a constant determined by the initial conditions.

If $f_0 < -1$, the initial value of the solution at $t = 0$ is

$$f(0) = \frac{f_0}{1 + f_0} = \frac{|f_0|}{|f_0| - 1} > 1 \tag{2.87}$$

and $\dot{f} < 0$, which makes this solution decrease and approach 1 asymptotically. If instead $f_0 > 0$, then it is $f(0) < 1$ and the sigmoid solution is such that $f < 1$. Here we impose that the solution f varies in the range $0 < f \leq 1$.

Another solution is the constant function $f \equiv 1$, which is a fixed point solution and a late–time attractor. Yet another constant solution $f(t) \equiv 0$ exists, and it is also an equilibrium point and a repellor, but it is irrelevant when we restrict to $f > 0$.

The logistic equation offers two analogies with FLRW cosmology, according the two different time coordinates used in cosmology, comoving and conformal time. As for the Fibonacci numbers discussed in the previous section, the analogy leads naturally to the introduction of a Lagrangian and a Hamiltonian for the logistic equation. Finally, the symmetries of the Friedmann equation can be used to highlight corresponding symmetries of the logistic equation.

The now familiar procedure of dividing the logistic equation (2.85) by f and squaring yields the differential equation

$$\left(\frac{\dot{f}}{f} \right)^2 = r^2 - 2r^2 f + r^2 f^2, \tag{2.88}$$

which can be regarded as the analogue of the Friedmann equation (1.179) [119].

2.5.1 *Comoving time analogy I*

By identifying the quantities

$$\left(t,\, a(t)\right) \leftrightarrow \left(t,\, f(t)\right), \tag{2.89}$$

$$K = 0, \tag{2.90}$$

$$\Lambda = 3r^2, \tag{2.91}$$

Eq. (2.88) becomes the analogue of the Friedmann equation (1.179) with energy density

$$\rho = \rho_0\, a\, (a-2) \tag{2.92}$$

and $\rho_0 = 3r^2/(8\pi G)$. While in FLRW cosmology the scale factor can be rescaled by a positive constant without affecting the physics, on the logistic side of the analogy the sigmoid solution $f(t)$ of the logistic equation cannot be rescaled in this way because the term $1-f$ appears in the right hand side of Eq. (2.85) and rescaling f would change this equation in an essential way. For a perfect physical analogy one should require that the energy density ρ is non–negative, which would force the lower bound $a(t) \geq 2$ $\forall t$ on the scale factor. However, if the interest lies only in the mathematical properties of the formal analogy, one can relax this assumption.

For the formal analogy with FLRW cosmology to be complete, another one of the Einstein–Friedmann equations must be satisfied. We impose the covariant conservation equation (1.187), which dictates the equation of state of the effective cosmic fluid

$$P = -\rho - \frac{a}{3} \frac{d\rho}{da}. \tag{2.93}$$

Let us focus on Eq. (2.92), viewing it as the quadratic algebraic equation

$$a^2 - 2a - \frac{\rho}{\rho_0} = 0. \tag{2.94}$$

Only the positive root

$$a = 1 + \sqrt{1 + \frac{\rho}{\rho_0}} \tag{2.95}$$

is relevant for cosmology hence we restrict to this, which is consistent with the assumption $f > 0$ in the logistic equation. Inserting Eq. (2.95) into Eq. (2.93) provides the equation of state of the analogous cosmic fluid

$$P = -\frac{5}{3}\rho - \frac{2\rho_0}{3}\left[1 + \sqrt{1 + \frac{\rho}{\rho_0}}\,\right], \tag{2.96}$$

which is non-linear and of phantom nature, *i.e.*, $P/\rho < -1$.

The universe analogous to the logistic equation is spatially flat, has positive cosmological constant Λ, and is permeated by a perfect fluid with the exotic equation of state (2.96). If the energy density were to be non–negative, the scale factor would be forced to satisfy $a \geq 2$, but the sigmoid (2.86) is never larger than unity. So, either one allows ρ to be negative or abandons this first analogy. If the interest is in the formal mathematical properties, this analogy remains viable.

The inverse variational problem can be posed for the logistic equation (2.85) and then the standard literature on FLRW cosmology suggests the Lagrangian for the logistic equation

$$L_1 \left(f, \dot{f} \right) = f \dot{f}^2 + r^2 f^4 \left(f - 2 \right) + r^2 f^3 . \tag{2.97}$$

Clearly, L_1 does not depend explicitly on time and the corresponding Hamiltonian

$$\mathcal{H}_1 = \frac{\partial L_1}{\partial \dot{f}} \dot{f} - L_1 = f \dot{f}^2 - r^2 f^4 \left(f - 2 \right) - r^2 f^3 = C_1 \tag{2.98}$$

is conserved. We must set the integration constant $C_1 = 0$, which produces the two equations

$$\dot{f} = \pm |r| f \left(1 - f \right) , \tag{2.99}$$

in which the negative sign corresponds to inverting the sign of the parameter r of the logistic equation, which is instead positive. We discard the negative sign.

The symmetries (1.228)–(1.229) and (1.230)–(1.232) do not apply to this analogous FLRW universe because the equation of state of the analogous cosmic fluid is non-linear. Also the third symmetry is not applicable because it applies to a Friedmann equation (1.179) with a single perfect fluid, possibly in the presence of the cosmological constant Λ [118], while here we have two fluids plus Λ. In fact, eliminating one of the fluids necessarily sets $r = 0$ and the logistic equation is lost. Therefore, also the third symmetry does not apply to Eq. (2.88).

2.5.2 *Comoving time analogy II*

A second analogy is generated if we take a spatially flat analogous universe with $\Lambda = 0$ and energy density

$$\rho = \rho_0 \left(1 - a \right)^2 . \tag{2.100}$$

To complete this analogy we impose the covariant energy conservation equation (1.187), which gives the equation of state of the analogous cosmic fluid

$$P = -\rho + \frac{2\rho_0}{3} a (1 - a) . \tag{2.101}$$

We can eliminate the scale factor a using Eq. (2.100), obtaining

$$a = 1 \pm \sqrt{\frac{\rho}{\rho_0}} \tag{2.102}$$

and the non-linear equation of state

$$P = -\frac{5}{3}\rho \mp \frac{2}{3}\sqrt{\rho_0\,\rho}. \tag{2.103}$$

The FLRW cosmos analogous to the logistic equation is spatially flat, has $\Lambda = 0$, is filled with a fluid with non–negative energy density and equation of state (2.103) and is accelerated, as follows from the acceleration equation (1.180) since $P < -\rho/3$.

The Lagrangian and Hamiltonian functions for the analogous cosmos are

$$L_2 = 3\,a\,\dot{a}^2 + 8\pi G\rho_0\,a^3\,(1 - a)^2 , \tag{2.104}$$

$$\mathcal{H}_2 = 3\,a\,\dot{a}^2 - 8\pi G\rho_0\,a^3\,(1 - a)^2 , \tag{2.105}$$

and reproduce the Lagrangian (2.97) and the Hamiltonian (2.98) for the logistic equation already seen, apart from an irrelevant constant.

Again, due to the non–linearity of the equation of state, the symmetries (1.228)–(1.229) and (1.230)–(1.232) do not apply. The third FLRW symmetry (1.233)–(1.235) respects the analogy and, therefore, preserves the logistic equation if the equation of state (2.103) is maintained. This is a non–trivial requirement which is satisfied if

$$\pm \bar{\rho}^{3/2} + \sqrt{\bar{\rho}_0}\,\bar{\rho} = \left(\pm\rho^{3/2} + \sqrt{\rho_0}\,\rho\right) \frac{d\bar{\rho}}{d\rho}, \tag{2.106}$$

which implies that

$$\int \frac{d\bar{\rho}}{\sqrt{\bar{\rho}_0}\,\bar{\rho} \pm \bar{\rho}^{3/2}} = \int \frac{d\rho}{\sqrt{\rho_0}\,\rho \pm \rho^{3/2}} . \tag{2.107}$$

The integral is computed explicitly,

$$\frac{2}{\sqrt{\bar{\rho}_0}} \ln\left(\frac{\sqrt{\bar{\rho}}}{\sqrt{\bar{\rho}} \pm \sqrt{\bar{\rho}_0}}\right) = \frac{2}{\sqrt{\rho_0}} \ln\left(\frac{\sqrt{\rho}}{\sqrt{\rho} \pm \sqrt{\rho_0}}\right) , \tag{2.108}$$

and we rewrite it as

$$\ln\left(1 \pm \sqrt{\frac{\bar{\rho}_0}{\bar{\rho}}}\right) = \sqrt{\frac{\bar{\rho}_0}{\rho_0}} \ln\left(1 \pm \sqrt{\frac{\rho_0}{\rho}}\right). \tag{2.109}$$

Noting the relation between densities and logistic parameters

$$\sqrt{\frac{\bar{\rho}_0}{\rho_0}} = \frac{\bar{r}}{r}, \tag{2.110}$$

we write

$$1 \pm \sqrt{\frac{\bar{\rho}_0}{\bar{\rho}}} = \left(1 \pm \sqrt{\frac{\rho_0}{\rho}}\right)^{\bar{r}/r} \tag{2.111}$$

and, finally,

$$1 \pm \sqrt{\frac{\rho_0}{\rho}} = \frac{1 \pm 1 - f}{1 - f}. \tag{2.112}$$

Let us consider the two sign possibilities separately. If we use the lower sign, Eq. (2.111) gives

$$-\frac{\bar{f}}{1 - \bar{f}} = \left(-\frac{f}{1 - f}\right)^{\bar{r}/r}. \tag{2.113}$$

The argument of the parenthesis in the right hand side is negative, hence the expression is defined only if $\bar{r}/r \equiv n$ is an integer. Further, when n is even, then the left hand side of Eq. (2.113) is negative while its right hand side is positive and Eq. (2.113) does not admit solutions. If n is odd, $n = 2k + 1$ with $k \in \mathbb{N}$, Eq. (2.113) becomes

$$f \to \bar{f} = \frac{\left(\frac{f}{1-f}\right)^{2k+1}}{1 + \left(\frac{f}{1-f}\right)^{2k+1}}. \tag{2.114}$$

Not surprisingly, the sigmoid solution (2.86) of the logistic equation exhibits this symmetry, which coincides with the rescaling

$$f = \frac{f_0\, e^{\beta t}}{1 + f_0\, e^{\beta t}} \to \bar{f} = \frac{\bar{f}_0\, e^{\bar{\beta} t}}{1 + \bar{f}_0\, e^{\bar{\beta} t}}, \tag{2.115}$$

$$\bar{f}_0 = f_0^{2k+1}, \tag{2.116}$$

$$\bar{\beta} = (2k + 1)\beta. \tag{2.117}$$

This scaling can be derived directly from the scaling properties of the exponential function appearing in the sigmoid solution.

The other possibility left consists of using the upper sign in Eq. (2.112) instead. In this case one has

$$\frac{2 - \bar{f}}{1 - \bar{f}} = \left(\frac{2 - f}{1 - f}\right)^{\bar{r}/r} \equiv \gamma\,, \tag{2.118}$$

$$\bar{f} = \frac{2 - \gamma}{1 - \gamma}\,, \tag{2.119}$$

which map the solution (2.86) into

$$\bar{f} = 1 + \frac{1}{1 - \left(2 + f_0 e^{\beta t}\right)^{\bar{r}/r}}\,. \tag{2.120}$$

These symmetries of the logistic equation and of its solutions are certainly not new, but they are somehow hidden and the analogy with FLRW cosmology makes them explicit. For example, a systematic search for symmetries does not unearth Eqs. (2.114) or Eq. (2.120).

2.5.3   *Conformal time analogy I*

Several problems in cosmology are approached using conformal time η instead of comoving time t. The two are related by $dt \equiv a\, d\eta$ [435]. Under the assumption that the universe is filled by a single perfect fluid with constant equation of state $P = w\,\rho$ and that $\Lambda = 0$, and combining the Friedmann equation (1.179) with the acceleration equation (1.180) written in terms of conformal time, one has

$$\frac{a_{\eta\eta}}{a} + (c - 1)\left(\frac{a_\eta}{a}\right)^2 + c\kappa = 0\,, \tag{2.121}$$

where

$$c = \frac{3w + 1}{2}\,. \tag{2.122}$$

The use of the new variable

$$u \equiv \frac{a_\eta}{a}\,, \tag{2.123}$$

turns Eq. (2.121) into the Riccati equation [132]

$$u_\eta + c\,u^2 + c\kappa = 0\,. \tag{2.124}$$

This procedure constitutes an alternative way of solving the Einstein–Friedmann equations in the presence of a single perfect fluid with constant equation of state [25, 132]. The reduction to a Riccati equation is the basis for several applications in cosmology [80, 199, 343–348, 361].

Given that the logistic equation is analogous to a Friedmann equation, it can also be reduced to a Riccati equation, and this fact is well known [361]. There are other integrable first order equations that are related with FLRW universes, especially those containing a scalar field minimally coupled to gravity, which is known to be equivalent to an effective perfect fluid. Just to quote a few examples, the solution of the Riccati equation (2.124) is a hyperbolic tangent, and cosmological models based on a phantom scalar field are known which are described by an hyperbolic tangent [9]. A cosmological model based on a scalar field with exponential potential (the trademark of dimensionally reduced theories) presented in Ref. [357] contains a first order equation used to integrate the model. Scalar field cosmology exhibits also a connection with the Abel equation of the first kind, discussed in Ref. [449]. Important connections exist between the Einstein–Friedmann equations and the Ermakov–Pinney equation [203] or the Schrödinger equation [18, 168].

Let us continue with our Riccati equation (2.124): we set

$$f(t) \equiv \frac{1}{2}\Big[g(t) + 1\Big] \tag{2.125}$$

turning Eq. (2.85) into

$$\dot{g} + \frac{r}{2}g^2 - \frac{r}{2} = 0\,, \tag{2.126}$$

which has the form (2.124) with $c = r/2$ and $\kappa = -1$. Thus, we have established the analogy

$$\Big(\eta, u(\eta)\Big) \to \Big(t, g(t)\Big) = \Big(t, 2f(t) - 1\Big)\,, \tag{2.127}$$

$$\kappa = -1\,, \tag{2.128}$$

$$c = \frac{r}{2} = \frac{3w + 1}{2}\,, \tag{2.129}$$

which gives

$$w = \frac{r - 1}{3}\,. \tag{2.130}$$

Two classic solutions of FLRW cosmology [82, 244, 260, 435] are obtained for the special cases $r = 1$ (dust) and $r = 2$ (radiation).

For this formal analogy to be complete, the equation

$$\frac{a_\eta}{a} = u = g(\eta) = 2f(\eta) - 1 \tag{2.131}$$

has to be satisfied and we can use the sigmoid solution (2.86) to integrate it and obtain the scale factor of the analogous universe

$$a(\eta) = a_0\, e^{-\eta} \left(1 + f_0 e^{r\eta}\right)^{2/r} \tag{2.132}$$

in conformal time η. The comoving time t can be obtained from the conformal time by integrating the relation $dt = a\, d\eta$ and is expressed by the hypergeometric function

$$t = \int a\, d\eta = -a_0\, e^{-\eta}\, {}_2F_1\left(-\frac{2}{r}, -\frac{1}{r}, \frac{r-1}{r}, -f_0\, e^{r\eta}\right) + t_0\,, \tag{2.133}$$

where t_0 is an integration constant. However, the parametric representation of the solution $a(t)$ in terms of comoving time (with η as the parameter) given by Eqs. (2.132) and (2.133) is too cumbersome for practical uses.

It is easy to guess a Lagrangian for the Riccati equation (2.126) in the form

$$L_R\left(g, \dot{g}\right) = \dot{g}^2 + \frac{r^2}{4}\left(g^2 - 1\right)^2\,. \tag{2.134}$$

This Lagrangian does not depend explicitly on η and the corresponding Hamiltonian

$$\mathcal{H}_R = \dot{g}^2 - \frac{r^2}{4}\left(g^2 - 1\right)^2 \tag{2.135}$$

is conserved. Its constant value must be set to zero and then the negative sign of the two roots of $\mathcal{H} = 0$ is chosen to reproduce Eq. (2.126).

One expects the symmetries (1.228)–(1.229) and (1.230)–(1.232) not to apply to the Riccati equation

$$u_\eta + c\,u^2 - c = 0 \tag{2.136}$$

because this implies $\kappa = -1$. However, the first symmetry becomes

$$u \to \bar{u} = \frac{1}{u} \tag{2.137}$$

and leaves the equation invariant. The second symmetry of the Einstein–Friedmann equations, instead, is not a symmetry of the Riccati equation. The third symmetry only applies to $\kappa = 0$ FLRW universes, while here $\kappa = -1$.

2.5.4 *Conformal time analogy II*

There is a second analogy $\left(\eta, a(\eta)\right) \to \left(t, f(t)\right)$ in addition to the one $\left(\eta, u(\eta)\right) \to \left(t, f(t)\right)$ already discussed.

The Friedmann equation written in terms of conformal time η has the form

$$\left(\frac{a_\eta}{a}\right)^2 = \frac{\Lambda a^2}{3} + \frac{8\pi G}{3}\rho a^2 - K, \qquad (2.138)$$

which is analogous to the logistic equation (2.88) through the correspondence

$$K = -r^2, \qquad (2.139)$$

$$\Lambda = 3\,r^2, \qquad (2.140)$$

$$\rho = -\frac{3\,r^2}{4\pi G a} \equiv -\frac{\rho_0}{a}. \qquad (2.141)$$

Again, the energy density (2.141) is negative, limiting the analogy to its more formal aspects.

By viewing the cosmological constant Λ as an effective perfect fluid with energy density $\rho_\Lambda = \Lambda/(8\pi G)$ in addition to the matter fluid, the Lagrangian familiar from FLRW cosmology is

$$L_3 = \frac{a_\eta^2}{a} + \frac{8\pi G}{3}\rho a^3 - K a + \frac{\Lambda a^2}{3}, \qquad (2.142)$$

while the associated Hamiltonian reads

$$\mathcal{H}_3 = \frac{a_\eta^2}{a} - \frac{8\pi G}{3}\rho\, a^3 + K a - \frac{\Lambda a^2}{3}. \qquad (2.143)$$

Under the map (2.139)–(2.141) the logistic equation is equivalent to $\mathcal{H}_3 = 0$ on the FLRW side of the analogy, provided that the positive sign is adopted when taking the square root of both sides.

The spatial sections of the analogous universe are spatially curved, hence the symmetries (1.228)–(1.229) and (1.230)–(1.232) do not apply. The third symmetry belongs to a universe filled by a single perfect fluid (possibly with Λ) [118]. However, eliminating the second (or κ–) fluid by setting $\kappa = 0$ means setting $r = 0$ and losing the logistic equation. As a result, the third symmetry is not applicable to Eq. (2.138).

To conclude, the analogies presented extend to the solutions of the logistic equation certain properties of the corresponding analytical solutions

of the Einstein–Friedmann equations. It is shown in Ref. [87] that all solutions of the Friedmann equation (1.179) are roulettes. A roulette is the
trajectory described, in two dimensions, by a point that lies on a curve
rolling without slipping on another given curve. By extension, all solutions
of the logistic equation are also roulettes.

A relevant question in FLRW cosmology is under which conditions analytical solutions of the Einstein–Friedmann equations can be obtained explicitly in terms of elementary functions. This question is answered in
Ref. [87] using Chebyshev's theorem of integration [85, 273]. Transposing
this result to the logistic equation does not generate useful new information
because the solution of the logistic is the well known sigmoid.

Effective Lagrangians and Hamiltonians for the logistic equation are
obtained through the cosmological analogy, with the consequence that a
system ruled by the logistic equation possesses a conserved energy, which
vanishes. The symmetries of the Einstein–Friedmann equations unearth
hidden symmetries of the logistic equation, although it is not obvious that
the latter have useful applications. Overall, this analogy is less useful than
others examined in this book, but viewing the same problem from a different
angle is *a priori* interesting.

2.6   Geodesics of the Euclidean plane

Moving on to other analogies, we examine one originating in a simple problem of geometry [145]. Geodesic curves are those that extremize the length
between two fixed points in a given geometric space and finding them constitutes a classic variational problem. The simplest space that one can
consider is the Euclidean plane and the corresponding variational problem
of finding its geodesic curves, which are well known to be straight lines,
is also simple. It is not obvious, however, that this variational problem,
and its associated Euler–Lagrange equation, gives rise to an analogy with
FLRW cosmology [145].

Let us begin with the infinitesimal arc length along a curve $y(x)$ in the
Euclidean plane, which is

$$d\ell = \sqrt{dx^2 + dy^2} = \sqrt{1 + y'^2}\, dx\,, \qquad (2.144)$$

where $y' \equiv dy/dx$. Then, the finite length between two fixed points 1 and 2
along the curve $y(x)$ is the functional of this curve

$$\ell_{12}\Big[y(x)\Big] = \int_1^2 d\ell = \int_{x_1}^{x_2} dx\, \sqrt{1 + y'^2} \equiv \int_{x_1}^{x_2} dx\, L\left(y'(x)\right)\,, \qquad (2.145)$$

where

$$L(y') = \sqrt{1 + y'^2} \qquad (2.146)$$

is the Lagrangian. Given that $\partial L/\partial y = 0$, the momentum

$$\pi_y = \frac{\partial L}{\partial y'} \qquad (2.147)$$

canonically conjugated to y is conserved,

$$\frac{y'}{\sqrt{1 + y'^2}} = C. \qquad (2.148)$$

By manipulating this equation one easily obtains

$$y' = \frac{C^2}{1 - C^2}, \qquad (2.149)$$

which has the linear solution

$$y(x) = \frac{C^2}{1 - C^2} x + D, \qquad (2.150)$$

i.e., the geodesics of the Euclidean plane are straight lines. We can re–arrange Eq. (2.150) as

$$\left(\frac{y'}{y}\right)^2 = \frac{C_1^2}{y^2} \qquad (2.151)$$

where

$$C_1^2 = \frac{C^2}{1 - C^2}. \qquad (2.152)$$

Equation (2.151) is the analogue of the very simple Friedmann equation (1.179)

$$H^2 = \frac{C_1^2}{a^2} \qquad (2.153)$$

corresponding to an empty universe with hyperbolic spatial sections. This is well known in cosmology as the "Milne universe" (*e.g.*, [288]). It is a trivial solution of the Einstein–Friedmann equations, the flat Minkowski spacetime, but it is disguised by accelerated coordinates that give a hyperbolic foliation of flat spacetime with negatively curved spatial slices of constant time. When computed, all the components of the curvature tensor vanish identically, but the intrinsic curvature of the spatial three–sections $t = \text{const.}$ does not vanish [288].

2.7   Geodesics of the Poincaré half–plane

Let us consider now a non–trivial two–dimensional space and its geodesics. The Poincaré half–plane consists of the upper part of the (x, y) plane with $y > 0$ equipped with the line element

$$d\ell^2 = \frac{1}{y^2} \left(dx^2 + dy^2 \right) ,$$
(2.154)

which is clearly conformal to the Euclidean line element $d\ell^2_{(0)} = dx^2 + dy^2$ in Cartesian coordinates. The infinitesimal arc length along a curve $y(x)$ is

$$d\ell = \frac{\sqrt{dx^2 + dy^2}}{y}$$
(2.155)

and the geodesic curves of this space are found through the corresponding variational principle. By extremizing the finite length along a curve $y(x)$

$$\ell_{12}\left[y(x) \right] = \int_{x_1}^{x_2} dx \, \frac{\sqrt{1 + y'^2}}{y} \equiv \int_{x_1}^{x_2} dx \, L\left(y, y' \right)$$
(2.156)

between two given points, one obtains the Lagrangian

$$L\left(y, y' \right) = \frac{\sqrt{1 + y'^2}}{y}$$
(2.157)

and the corresponding Euler–Lagrange equation. Instead of solving the second order Euler–Lagrange equation, we note that $\partial L/\partial x = 0$ and, therefore, the Hamiltonian is conserved. This gives

$$\frac{\partial L}{\partial y'} y' - L = C < 0 ,$$
(2.158)

which leads to the first order differential equation

$$\frac{1}{Cy} = -\sqrt{1 + y'^2} .$$
(2.159)

The solutions for $C = -1$ are half–circles perpendicular to the x–axis.

There is a non–trivial cosmological analogy for this equation, which is discussed in Rindler's textbook [342] and in Refs. [94,145]. The corresponding Friedmann equation

$$H^2 = \frac{1}{C^2 a^4} - \frac{1}{a^2}$$
(2.160)

describes a positively curved universe permeated by blackbody radiation with energy density

$$\rho = \frac{\rho_0}{a^4},$$ (2.161)

$$\rho_0 = \frac{3}{8\pi GC^2},$$ (2.162)

and pressure $P = \rho/3$. The corresponding scale factor is the textbook solution of the Einstein–Friedmann equation [435],

$$a(t) = \sqrt{C'}\sqrt{1 - \left(1 - \frac{t}{C'}\right)^2},$$ (2.163)

with $C' = 1/C^2$.

2.8 Cosmology and the catenary problem

The catenary problem is a classic problem of variational calculus. If a heavy rope is hang between two points, it assumes the shape of a catenary (Figs. 2.2 and 2.3).

Consider a heavy string suspended in a vertical (x, y) plane. Its configuration of equilibrium is described by a profile $y(x)$ that we want to determine. Let the linear mass density of the string be $\mu = dm/d\ell$, where

$$d\ell = \sqrt{dx^2 + dy^2} = \sqrt{1 + (y')^2}\, dx$$ (2.164)

is the infinitesimal arc length along the string profile. An infinitesimal element of the string with length $d\ell$ at the horizontal coordinate x has gravitational potential energy

$$dE_g = \mu\, g\, y(x)\, d\ell,$$ (2.165)

where g is the acceleration of gravity. The total gravitational potential energy of this string suspended by two points of horizontal coordinates x_1 and x_2 is obtained by integrating between these points,

$$E_g\Big[y(x)\Big] = \mu\, g \int_{x_1}^{x_2} dx\, y\, \sqrt{1 + (y')^2} \equiv \int_{x_1}^{x_2} L\, dx,$$ (2.166)

and produces a functional of the curve $y(x)$, where the Lagrangian is (apart from the constant $\mu\, g$)

$$L(y, y') = y\sqrt{1 + (y')^2}.$$ (2.167)

Fig. 2.2. A heavy rope hung between two points takes the shape of a catenary.

This Lagrangian does not depend explicitly on the coordinate x and the corresponding Hamiltonian is conserved [183]

$$\mathcal{H} = \frac{\partial L}{\partial y'} y' - L = c_1 , \qquad (2.168)$$

where c_1 is an integration constant. This equation (Beltrami identity) can be written as

$$\frac{-y}{\sqrt{1 + (y')^2}} = c_1 , \qquad (2.169)$$

and its solutions are catenaries [183].

An extra ingredient which, however, is not necessary consists of requiring that the string has fixed length between x_1 and x_2, which is implemented as a Lagrangian constraint in the variational integral. The new variational integral is

$$J\Big[y(x)\Big] = \int_{x_1}^{x_2} dx \, (y + \lambda) \sqrt{1 + (y')^2} , \qquad (2.170)$$

where λ is a Lagrange multiplier. The shift of the y–coordinate $y \to \bar{y} \equiv y + \lambda$ then reduces this problem to the previous one.

FIG. 2.3. Heavy dew makes each arc of this spider web sag in the shape of a catenary.

Equation (2.169) has as a cosmological analogue a well known scenario of FLRW cosmology [145]: by re–arranging (2.169) as

$$\left(\frac{y'}{y}\right)^2 = \frac{1}{C_1^2} - \frac{1}{y^2} \, ,\qquad (2.171)$$

one can read the analogous Friedmann equation

$$H^2 = \frac{\Lambda}{3} - \frac{1}{a^2} \qquad (2.172)$$

for the scale factor $a(t)$, where $\Lambda = 3/C_1^2 > 0$ is the cosmological constant. The analogous universe is the empty $\kappa = +1$ solution of the Einstein–Friedmann equations with $\Lambda > 0$. The scale factor is the catenary

$$a(t) = \sqrt{\frac{3}{\Lambda}} \, \cosh\left[\sqrt{\frac{\Lambda}{3}} \, (t - t_0)\right] . \qquad (2.173)$$

This universe contracts from infinite size in the past $t < t_0$, bounces at the minimum size $\sqrt{3/\Lambda}$ at $t = t_0$ thanks to the cosmological constant (which avoids a spacetime singularity $a = 0$ by violating the strong energy condition), and then expands forever for $t > t_0$, asymptoting to the de

Sitter space with scale factor

$$a(t) = \sqrt{\frac{3}{4\Lambda}} \exp\left[\sqrt{\frac{\Lambda}{3}}\,(t - t_0)\right] \tag{2.174}$$

as $t \to +\infty$. The curvature becomes less and less relevant in this limit.

2.9   Cosmic analogue of the minimal surface of revolution

Finding the minimal surface of revolution constitutes another textbook problem of variational calculus. Let $y(x)$ be the equation of a curve that joins two points in the vertical (x, y) plane. Rotate this curve about the vertical axis and pose the question of which curve $y(x)$ minimizes the lateral area spanned. This problem is solved by computing the area integral [183]

$$J = \int_{x_1}^{x_2} dx\, x\, \sqrt{1 + \left(\frac{dy}{dx}\right)^2}, \tag{2.175}$$

which is a functional of the curve $y(x)$, and by extremizing it. As a practical application, and for visualization, one can think of soapy membranes between wire frames. The energy of a membrane of soapy water is proportional to its area, therefore a membrane left to itself tends to minimize its surface area to reach the configuration of minimum energy. If soapy water is placed on a wire frame obtained by rotating a wire shaped as the graph of $y(x)$ about the y–axis, it arranges itself in a configuration of equilibrium that corresponds to the minimal surface of revolution that has the wired frame as its boundary. The discussion can proceed in two different ways, according to whether x or y is taken as the independent variable.

2.9.1   *Dependent variable $x = x(y)$*

Suppose that we choose y as the independent variable and $x(y)$ as the dependent one. The variational integral is then rewritten as

$$J\left[x(y)\right] = \int_{y_1}^{y_2} dy\, x\, \sqrt{1 + \left(\frac{dx}{dy}\right)^2} \equiv \int_{y_1}^{y_2} dy\, L\left(x(y), \frac{dx}{dy}\right), \tag{2.176}$$

where $y_{1,2} \equiv y(x_{1,2})$. We do not need to write down, and solve, the second order Euler–Lagrange equation because $\partial L / \partial y = 0$ and the corresponding Hamiltonian is conserved. We have

$$\mathcal{H} = \frac{\partial L}{\partial x'}\, x' - L = \frac{x\, x'^2}{\sqrt{1 + x'^2}} - x\, \sqrt{1 + x'^2} = C, \tag{2.177}$$

where C is an integration constant and $x' \equiv dx/dy$. This equation can be rearranged to give

$$x'^2 = \frac{x^2}{C^2} - 1. \tag{2.178}$$

Separation of variables yields

$$\int \frac{dx}{\sqrt{x^2 - C^2}} = \ln\left(\sqrt{x^2 - C^2} + x\right) = \frac{y - y_0}{C}, \tag{2.179}$$

with y_0 another integration constant. Exponentiating both sides and manipulating yields

$$x(y) = \frac{1}{2}\left(e^{\frac{y-y_0}{C}} + C^2 e^{-\frac{y-y_0}{C}}\right). \tag{2.180}$$

The catenary curve

$$x(y) = \cosh\left(y - y_0\right) \tag{2.181}$$

is obtained with the initial conditions $C = \pm 1$, where $x_0 \equiv x(0)$ and $y_0 = \cosh^{-1} x_0$.

Let us come to the cosmic analogy, which is obtained by writing (2.178) as the equation

$$\left(\frac{x'}{x}\right)^2 = \frac{1}{C^2} - \frac{1}{x^2} \tag{2.182}$$

analogous to the Friedmann equation (1.179) in the special implementation

$$H^2 = \frac{\Lambda}{3} - \frac{1}{a^2}. \tag{2.183}$$

The analogous universe is positively curved, with curvature term $-K/a^2$, and has a positive cosmological constant $\Lambda = 3/C^2$; it coincides with the analogous universe of the catenary problem discussed in the previous section.

2.9.2 Dependent variable $y = y(x)$

Let us examine now the other choice of variables, in which x and y are the independent and the dependent variables, respectively. The Lagrangian is now

$$L\left(x, y'(x)\right) = x\sqrt{1 + y'^2}. \tag{2.184}$$

Again, we do not need the second order Euler–Lagrange equation, but only a first integral. $\partial L / \partial y = 0$ and the conjugate momentum is conserved, giving

$$\frac{\partial L}{\partial y'} = \frac{x \, y'}{\sqrt{1 + y'^2}} = C \,, \tag{2.185}$$

or

$$y' = \frac{C}{\sqrt{x^2 - C^2}} \,. \tag{2.186}$$

The integration is trivial, providing the solution

$$y(x) = C \ln \left(\sqrt{x^2 - C^2} + x \right) + \text{const.} \,, \tag{2.187}$$

which can be inverted and gives again

$$x(y) = \cosh (y - y_0) \tag{2.188}$$

if $C = \pm 1$. This time there is no cosmic analogy because the would–be analogue of the Friedmann equation

$$\frac{y'^2}{y^2} = \frac{C^2}{y^2 \left(x^2 - C^2 \right)} \tag{2.189}$$

contains explicitly the independent variable x. By contrast, the Friedmann equation (1.179) does not contain t explicitly.

2.10   Strategies to find analogies

Let us summarize some key points that are useful in the search for analogies between FLRW cosmology and other systems.

One can look for one–dimensional problems that have roulettes as solutions. In fact, all the solutions of the Friedmann equation are roulettes [87] (remember that a roulette is the curve described by a point attached to a closed convex curve as that curve rolls without slipping along a second, given, curve).

It is possible that a first order ordinary differential equation does not have a cosmological analogue but, upon a change of variables, it does (an example is the Vialov equation of glaciology discussed in Sec. 6.4). However, if the variables and their interpretation are changed radically, this may no longer be a meaningful analogy.

All Lagrangians containing only kinetic energy are similar to the one discussed in the analogy for Omori's law in Sec. 7.3. If the coefficients of this generalized kinetic energy are not constant, then there is no analogue using these variables. If these coefficients are constant, or can be brought to constant by redefining the independent variable (which could be equivalent to using conformal time), then the analogous universe is the de Sitter one.

In retrospect, we can list some general rules for an analogy to work when we start from the equation

$$\left(\frac{\dot{y}}{y}\right)^2 = P(y) \tag{2.190}$$

and $P(y)$ is a polynomial of powers or inverse powers of y. (This is not the only viable possibility, of course, but alternatives require to consider non-linear equations of state of the analogous cosmic fluid, which are more exotic than the standard ones found in cosmology textbooks.)

- If the right hand side $P(y)$ is a polynomial with all terms of the same sign, the analogy will contain only fluids with positive energy densities. In addition, one could also have a cosmological constant $\Lambda > 0$ or negative spatial curvature in the analogous universe.
- If the right hand side polynomial $P(y)$ contains a negative term, the analogy does not involve negative densities only if this negative term is constant (corresponding to a cosmological constant $\Lambda < 0$) or scales as $1/y^2$, in which case it describes positive spatial curvature. In any other case, there is a fluid with negative energy density, which makes the analogy purely formal but still valuable to solve the inverse variational problem and to find symmetries.

To conclude, although the Friedmann equation (1.179) has the form of an energy conservation equation for a particle in one–dimensional motion under a conservative force, not all one–dimensional systems lend themselves to cosmological analogies because the covariant conservation equation (1.187) may not be satisfied, or because the fluid filling the analogous universe has a completely unphysical equation of state (for example, a negative energy density). Overall, even if one–dimensional systems with approximately conserved energies are common, cosmic analogies are not and should be valued when encountered.

In the following, we prioritize analogies in which the cosmic fluids have positive energy densities, we discuss the physics on both sides of the analogy, and we find Lagrangians and Hamiltonians in order to solve the corresponding inverse variational (or Helmoltz) problems. Furthermore, when the analogous universe is spatially flat, we can search for symmetries of the system analogous to FLRW symmetries.

Chapter 3

Heating, cooling, and their cosmic analogues

> *It is probably no exaggeration to say that all of theoretical physics proceeds by analogy.*

> Jeremy Bernstein

3.1 Physics of heating and cooling

There are three possible mechanisms of heat transfer: conduction, convection, and radiation. Conduction is described by the heat, or diffusion, equation which is a partial differential equation and does not lend itself to analogies with the Einstein–Friedmann equations (1.179)–(1.187), which are instead ordinary differential equations. Convection and radiation are more complicated physical phenomena, but they can be described with simplified models involving ordinary differential equations, which do admit analogies with FLRW cosmology.

Convection has been described for centuries. Newton proposed a simplified model of heating and cooling by convection already in 1701. Let T be the surface temperature of a cooling body (which, contrary to the temperature in blackbody theory, needs not be the absolute temperature of the Kelvin scale), Q the heat energy transferred, and A the surface area of the body. In Newton's model of cooling, the rate of convective heat loss per unit area is

$$\frac{1}{A}\frac{dQ}{dt} = -h\left(T - T_\infty\right) , \tag{3.1}$$

where T_∞ is the temperature of the surroundings and h is a convection coefficient, assumed to be constant for simplicity. The regime of validity of Newton's law of cooling (3.1) is limited: this law works well only for moderate temperature differences

$$\theta(t) \equiv T(t) - T_\infty .\tag{3.2}$$

Moreover, it works better to describe forced, rather than natural, convection (see the pedagogical review in Ref. [311]).

Let C be the (constant) heat capacity of the cooling body; then, the infinitesimal amount of heat transferred in the infinitesimal time dt causing the infinitesimal temperature change dT is $dQ = C dT$, which gives

$$\frac{dT}{dt} = -k\,(T - T_\infty)\tag{3.3}$$

where $k \equiv Ah/C$. For convenience, we use the temperature difference $\theta(t)$ instead of T, and then

$$\dot\theta = -k\,\theta .\tag{3.4}$$

The task of solving this cooling problem is simplified if the heat capacity of the surroundings is much larger than that of the cooling body: then T_∞ can be taken to be constant and the solution of Eq. (3.3) is simply

$$T(t) = (T_0 - T_\infty)\,e^{-kt} + T_\infty ,\tag{3.5}$$

with $T_0 = T(0)$ the initial condition on the temperature. As time goes by, the state of thermal equilibrium $T(t) \simeq T_\infty$ is approached, irrespective of the initial condition T_0.

When the temperature difference θ is large, other models of convective cooling are more appropriate. These include the 1817 Dulong–Petit law [116]

$$\frac{1}{A}\frac{dQ}{dt} = -g\Big(T - T_\infty\Big)^n ,\tag{3.6}$$

with the exponent in the range $1.25 \le n \le 1.6$ [69, 292, 311, 400]. In terms of the variable θ, we have

$$\frac{d\theta}{dt} = -g_0\,\theta^n\tag{3.7}$$

where $g_0 \equiv gA/C$. The solution of this first order ordinary differential equation is

$$\theta(t) = \frac{\theta_0}{\left| t - t_0 \right|^{\frac{1}{n-1}}} ,\tag{3.8}$$

where

$$\theta_0 = \left[(n-1)g_0 \right]^{\frac{1}{1-n}} \qquad (3.9)$$

and t_0 is an integration constant. This solution diverges as $t \to t_0$ and is not valid arbitrarily in the far past.

The description of cooling by radiation is dear to theoretical physicists because it is connected with the Planck blackbody distribution and the birth of quantum mechanics. The Stefan–Boltzmann law describes radiation with much better accuracy than phenomenological models describe convection. However, in everyday situations, convective cooling and radiative cooling often co–exist, resulting in the combined Newton–Stefan cooling law [311]

$$\frac{1}{A}\frac{dQ}{dt} = -h\left(T - T_\infty\right) - \epsilon\sigma\left(T^4 - T_\infty^4\right), \qquad (3.10)$$

where σ is the Stefan–Boltzmann constant, ϵ is the emissivity of the cooling body, and T is the Kelvin temperature. The ordinary differential equation ruling the temperature difference is now

$$\frac{d\theta}{dt} = -\left(\alpha\,\theta + \beta\,\theta^2 + \gamma\,\theta^3 + \delta\,\theta^4\right), \qquad (3.11)$$

where

$$\alpha = \frac{A}{C}\left(h + 4\,\epsilon\,\sigma\,T_\infty^3\right), \qquad (3.12)$$

$$\beta = \frac{6A}{C}\,\epsilon\,\sigma\,T_\infty^2, \qquad (3.13)$$

$$\gamma = \frac{4A\,\epsilon\,\sigma\,T_\infty}{C}, \qquad (3.14)$$

$$\delta = \frac{A\,\epsilon\,\sigma}{C}. \qquad (3.15)$$

Although Eq. (3.11) is rather complicated, it simplifies in the physically meaningful approximation in which the temperature difference θ is small. Then, the higher order terms in θ can be dropped [311] and one obtains the simplified evolution equation

$$\frac{d\theta}{dt} = \left(\alpha + \beta\,\theta\right)\theta. \qquad (3.16)$$

The general solution is

$$\theta(t) = \frac{\theta_0\,e^{-\alpha t}}{1 + \Gamma\,\theta_0\left(1 - e^{-\alpha t}\right)}, \qquad (3.17)$$

where $\theta_0 = \theta(0)$ is the initial temperature difference between the body and its surroundings and $\Gamma = \beta/\alpha$.

If we wait long enough, the temperature difference decreases, higher order terms in the fourth order polynomial in θ in Eq. (3.11) become negligible, the linear and quadratic terms come to dominate, and the simplified model (3.16) approximates the exact Newton–Stefan law better and better (the smoothing of temperature gradients is a characteristic feature of spontaneous heat transfer in the absence of steady sources). If we wait even longer, the Newton law of cooling incorporating only the linear term becomes realistic. The mathematical description of this physics is that the decreasing exponential solution $\theta_0\,e^{-\alpha t}$ of Newton's model constitutes a late–time attractor in the phase space of the solutions of the full Newton–Stefan cooling equation (3.11). Mathematically, this is the reason why the simplified model (3.16) makes sense.

Formal analogies between the various cooling laws and FLRW universes exist [156].

3.2   Cool cosmic analogies

Let us examine the cosmic analogues of the first order cooling laws and the Lagrangians, Hamiltonians, and symmetries that come with them. By now we are familiar with the inverse variational problem and with the fact that, although the Euler–Lagrange equation

$$\frac{d}{dt}\left(\frac{\partial L}{\partial \dot\theta}\right) - \frac{\partial L}{\partial \theta} = 0 \tag{3.18}$$

admits a first integral containing an arbitrary integration constant, only one specific value of this constant reproduces the original cooling equation.

3.2.1   *Newton cooling*

As expected, the case of Newton's law of cooling is almost trivial. Following our usual procedure, we divide Newton's law of cooling (3.3) by θ and square the result to obtain

$$\left(\frac{1}{\theta}\frac{d\theta}{dt}\right)^2 = k^2 . \tag{3.19}$$

This equation has the form of the Friedmann equation

$$H^2 = \frac{\Lambda}{3} \tag{3.20}$$

describing an empty $\kappa = 0$ universe with positive cosmological constant $\Lambda = 3\,k^2$. As is well known, the solutions are expanding or contracting de Sitter universes with scale factors

$$a(t) = a_0\,\mathrm{e}^{\pm kt} \tag{3.21}$$

where, for both cooling and heating, only the negative sign applies to the analogy. The physical temperature difference in the corresponding problem is $\theta(t) = \theta_0\,\mathrm{e}^{-kt}$.

The FLRW Lagrangian suggests the effective Lagrangian for Newton's law of cooling

$$L_1\left(\theta, \dot\theta\right) = \dot\theta^2 + \left(\frac{Ah}{C}\right)^2 \theta^2\,. \tag{3.22}$$

L_1 does not depend explicitly on time and the corresponding Hamiltonian is conserved [183],

$$\mathcal{H} = \dot\theta^2 - k^2\theta^2 = C\,. \tag{3.23}$$

The specific value $C = 0$ of the integration constant and the negative sign of the square root of the resulting equation give back Newton's law of cooling.

In this highly symmetric picture, the FLRW symmetry (1.230)–(1.232) is trivial and just rescales the units of time $t \to \bar t = s\,t$ and the scale factor $a \to \bar a = a_0^s\,\mathrm{e}^{-k\bar t}$, but does not alter the energy density: $\bar\rho = \rho$.

The other symmetry is also trivial. Since the cosmological constant is formally equivalent to a perfect fluid with a constant density $\rho_\Lambda = \Lambda/(8\pi G)$ and pressure $P_\Lambda = -\rho_\Lambda$, the symmetry (1.233)–(1.235) merely rescales Λ and the Hubble constant

$$\Lambda \to \bar\Lambda = \alpha\,\Lambda\,, \tag{3.24}$$

$$H = \sqrt{\frac{\Lambda}{3}} \to \bar H = \sqrt{\frac{\bar\Lambda}{3}} = \sqrt{\alpha}\,H\,. \tag{3.25}$$

3.2.2 Dulong–Petit cooling

An analogy with a different universe arises for the Dulong–Petit cooling law (3.7). In this case the analogue of the Friedmann equation reads

$$\left(\frac{1}{\theta}\frac{d\theta}{dt}\right)^2 = g_0^2\,\theta^{2(n-1)} \tag{3.26}$$

which is compared with

$$H^2 = \frac{8\pi G}{3}\,\rho\,. \tag{3.27}$$

The corresponding universe is again spatially flat but now $\Lambda = 0$ and the dynamics is dominated by a perfect fluid with equation of state $P = w\,\rho$ with

$$w = -\frac{(2n+1)}{3} \tag{3.28}$$

and

$$\rho_0 = \frac{3\,g_0^2}{8\pi G}. \tag{3.29}$$

Since $w < -1$, the equation of state is of the phantom type, causing the universe to end its history in a Big Rip singularity at a finite future. Phantom fluids [72, 73] (if they exist) constitute extremely exotic forms of dark energy and are often invoked to explain the present acceleration of the cosmic expansion [5]. From time to time, there are observational claims that the phantom equation of state is preferred over "regular" dark energy with $w \geq -1$. Since both pressure and energy density gravitate, for a phantom fluid the pressure dominates and makes the universe accelerate at such a rate that that it "explodes" at a finite time (Big Rip). In the Big Bang and Big Crunch singularities familiar from cosmology textbooks [82, 244, 260, 435], the scale factor $a(t)$ vanishes but at a Big Rip it becomes infinite, together with the scalar curvature invariants, the energy density, and the pressure. The expanding and contracting branches of the scale factor $a(t)$ are disconnected at the Big Rip: spacetime "ends" there.

Given the negative sign in the right–hand side of the cooling law, the universe mocking the cooling body contracts away from the Big Rip, as described by the scale factor

$$a(t) = \frac{a_0}{t - t_0} \tag{3.30}$$

for $t > t_0$. In Chap. 7 we discuss an analogy between the Omori–Utsu law of seismology [305, 411] and FLRW cosmology. The Omori–Utsu law describes phenomenologically the frequency $\dot{n}$ of the aftershocks following a main earthquake shock as

$$\dot{n} = -k\,n^p, \tag{3.31}$$

where k and p are constants. Therefore, the function $n(t)$ obeys the same formal equation as the scale factor of the universe in a Big Rip [146]. As a consequence, there exists an analogy between the Dulong–Petit law of cooling and the so–called "cooling" of active faults following a main earthquake shock. The expressions "cooling of a seismically active zone" and "hot zone" used in seismology acquire a precise meaning through the analogy with convective cooling. The dissipation of energy through secondary shocks is analogous to the convective heat loss from a hot body.

The power–law behaviour is ubiquitous in science and appears in physics, chemistry, astronomy, the earth sciences, biology, population dynamics, computer science, economics, information theory, language, and economics (*e.g.*, [331]). To quote another example, the recession flow of rivers in hydrology is described by the Brutsaert–Nieber law [64]

$$\frac{dQ}{dt} = -k\,Q^{\alpha}\,,\tag{3.32}$$

where Q is is the discharge at the river cross section at time t.

Let us discuss the inverse variational problem for Dulong–Petit cooling. Based on the experience with FLRW, we can easily identify a Lagrangian for the Dulong–Petit cooling process [156],

$$L_2\left(\theta,\dot\theta\right) = \theta\,\dot\theta^2 + g_0^2\,\theta^{2n+1}\tag{3.33}$$

together with the corresponding Hamiltonian

$$\mathcal{H} = \theta\,\dot\theta^2 - g_0^2\,\theta^{2n+1}\,,\tag{3.34}$$

which is conserved since $\partial L_2/\partial t = 0$. Choosing zero value for the integration constant appearing in this first integral and the negative sign in the square root gives back the Dulong–Petit cooling law.

Other Lagrangians are possible: for example

$$L_2' = \frac{1}{2}\left(c+t\right)^{\frac{n}{n-1}}\dot\theta^2\,,\tag{3.35}$$

which is explicitly time–dependent, is another Lagrangian [177]. Noting that $\partial L_2'/\partial\theta = 0$, one concludes that the momentum conjugated to θ

$$p_\theta \equiv \frac{\partial L_2'}{\partial\dot\theta} = \left(c+t\right)^{\frac{n}{n-1}}\dot\theta = c_0\tag{3.36}$$

(where c_0 is an integration constant) is conserved. The solution of this equation is

$$\theta(t) = \frac{-c_0\left(n-1\right)}{\left(c+t\right)^{\frac{1}{n-1}}}\,,\tag{3.37}$$

giving

$$\frac{1}{\left(c+t\right)^{\frac{1}{n-1}}} = \frac{\theta}{c_0\left(1-n\right)}\,.\tag{3.38}$$

Substituting into Eq. (3.36) yields

$$\dot{\theta} = \frac{c_0^{1-n}}{\left(1-n\right)^n} \, \theta^n \equiv -g_0^2 \, \theta^n \,. \tag{3.39}$$

It is interesting to look at the FLRW symmetries and their analogues. The symmetry operation (1.230)–(1.232) has a cousin in the Dulong–Petit law, expressed by

$$\theta \to \bar{\theta} = \theta^s \,, \tag{3.40}$$

$$dt \to d\bar{t} = s \, \theta^{(1-n)(s-1)} \, dt \,. \tag{3.41}$$

Scaling considerations are widely used in engineering and industrial applications of physics and this particular scaling invariance may be useful for such purposes.

Regarding the other symmetry (1.233)–(1.235), one notes that the equation of state of the analogous cosmic fluid must be preserved in order to preserve the analogy, with

$$\bar{w} = -\frac{(2\bar{n}+1)}{3} \,, \tag{3.42}$$

yielding

$$\left(\frac{\bar{\rho}}{\rho}\right)^{-3/2} \frac{d\bar{\rho}}{d\rho} = \frac{\bar{w}+1}{w+1} \,, \tag{3.43}$$

or

$$d\left(\bar{\rho}^{-1/2}\right) = \frac{\bar{w}+1}{w+1} \, d\left(\rho^{-1/2}\right) \,. \tag{3.44}$$

This first order differential equation is solved by

$$\bar{\rho} = \left(\frac{w+1}{\bar{w}+1}\right)^2 \rho = \left(\frac{n-1}{\bar{n}-1}\right)^2 \rho \tag{3.45}$$

and then it must be

$$\bar{\rho}_0 = \left(\frac{n-1}{\bar{n}-1}\right)^2 \rho_0 \,, \tag{3.46}$$

or

$$\bar{g}_0 = \frac{n-1}{\bar{n}-1} \, g_0 \,. \tag{3.47}$$

The symmetry equation (1.232) then gives

$$\bar{H} = \frac{n-1}{\bar{n}-1} H \tag{3.48}$$

and

$$\bar{a}(t) = \bar{a}_0 \, t^{\frac{n-1}{(1-n)(\bar{n}-1)}} = \left[a(t) \right]^{\frac{n-1}{\bar{n}-1}} . \tag{3.49}$$

To conclude, the temperature difference transforms according to

$$\theta \to \bar{\theta} = \theta^{\frac{n-1}{\bar{n}-1}} . \tag{3.50}$$

3.2.3 Simplified Newton–Stefan cooling

Let us consider another cooling model, the approximated Newton–Stefan cooling law (3.16). The usual division by θ and squaring give the analogue of the Friedmann equation

$$\left(\frac{1}{\theta} \frac{d\theta}{dt} \right)^2 = \alpha^2 + 2\alpha\beta\theta + \beta^2\theta^2 . \tag{3.51}$$

The mock universe is spatially flat, has positive cosmological constant $\Lambda = 3\alpha^2$, and contains two phantom fluids with energy densities

$$\rho_{(1)} = \frac{3\alpha\beta}{4\pi G} a , \tag{3.52}$$

$$\rho_{(2)} = \frac{3\beta^2}{8\pi G} a^2 , \tag{3.53}$$

and equation of state parameters $w_1 = -4/3$, $w_2 = -5/3$, respectively. Here we have $\dot{a} < 0$ and this analogous universe contracts. At late times, the solution (3.17) approaches

$$a(t) \simeq \sqrt{\frac{3}{\Lambda}} \frac{\theta_0 \, e^{-\sqrt{\Lambda/3}\,t}}{\sqrt{\Lambda/3} + \beta\theta_0} . \tag{3.54}$$

This property corresponds to the fact that the energy densities of the phantom fluids decay (scaling as a and a^2, respectively), while the cosmological constant with $\rho_\Lambda = \Lambda/(8\pi G)$ dominates at late times as the contracting universe evolves. The analogue of the contracting de Sitter space is a phase space attractor for the solutions of the simplified Newton–Stefan cooling model (indeed, it is an attractor also for the *exact* Newton–Stefan model since higher order powers of θ in the right–hand side of Eq. (3.11) decay faster than the lower order powers retained in the approximation (3.16)).

We can again introduce the Lagrangian

$$L_3\left(\theta, \dot{\theta}\right) = \theta\,\dot{\theta}^2 + \theta^3\left(\alpha + \beta\,\theta\right)^2 . \tag{3.55}$$

Setting $\beta = 0$ reproduces the Lagrangian

$$L_3\Big|_{\beta=0} = \theta\,L_1 , \tag{3.56}$$

which is equivalent to (3.22) describing Newton cooling. The other special case $\alpha = 0$ gives the Lagrangian (3.33) describing $n = 2$ Dulong–Petit cooling.

The new Lagrangian L_3 does not depend explicitly on the time t, hence the Hamiltonian

$$\mathcal{H} = \frac{\partial L_3}{\partial \dot{\theta}}\,\dot{\theta} - L_3 = \theta\,\dot{\theta}^2 - \theta^3\left(\alpha + \beta\,\theta\right)^2 = C_0 \tag{3.57}$$

(where C_0 is an integration constant) is conserved. Then,

$$\left(\frac{1}{\theta}\frac{d\theta}{dt}\right)^2 = \left(\alpha + \beta\,\theta\right)^2 + \frac{C_0}{\theta^3} \tag{3.58}$$

and the special value $C_0 = 0$ yields

$$\dot{\theta} = \pm\left(\alpha\,\theta + \beta\,\theta^2\right). \tag{3.59}$$

As usual, only the lower sign reproduces the original physical problem. Since the analogous universe contains two perfect fluids, in addition to Λ, the symmetries (1.230)–(1.235) of a spatially flat universe with a single fluid cannot be applied.

There are connections with other fields of mathematics and physics as well. If the signs of the coefficients are inverted, the equation describing the truncated Newton–Stefan model reproduces known equations of other areas of science. For example, one obtains the logistic equation discussed in Chap. 2 with its FLRW analogies. Another example is a simplified model of laser emission, in which the photon emission rate dn/dt is related to the number of photons $n(t)$ present in an excited state by [194, 361]

$$\frac{dn}{dt} = -\alpha\,n - \beta\,n^2 . \tag{3.60}$$

3.2.4   *Other cooling models*

One can consider more general models of convective–radiative cooling, for example

$$\frac{d\theta}{dt} = -\theta^p\left(\alpha + \beta\,\theta^r\right)^n , \tag{3.61}$$

where $n > 0$ is an integer. To integrate this first order differential equation one must compute the integral

$$\int d\theta \, \theta^{-p} \left(\alpha + \beta \, \theta^r \right)^{-n} , \tag{3.62}$$

which, provided that r is rational, has the form appearing in the Chebyshev theorem [85, 273]. No thermal physics experiment has the power to discriminate between a real number r and its rational approximation, so we can assume that r is always rational. Since $q = -n$, the Chebyshev theorem applies and the solution of Eq. (3.61) can always be expressed in terms of elementary functions. The integral in the equation

$$\int d\theta \, \theta^{-p} \left(\alpha + \beta \, \theta^r \right)^{-n} = - (t - t_0) , \tag{3.63}$$

where t_0 is an integration constant, can always be computed in terms of elementary functions, although it may not be possible to invert explicitly the function $t(\theta)$. For example, in the cooling model with parameters $p = 2, r = 1$, and $n = 2$ one has

$$\frac{\alpha \, \beta}{\alpha + \beta \, \theta} + \frac{\alpha}{\theta} + 2\beta \ln \left(\frac{\theta}{\alpha + \beta \, \theta} \right) = \alpha^3 \, (t - t_0) . \tag{3.64}$$

In all cooling models with a FLRW cosmological analogue, the graph of the solution $\theta(t)$ is a roulette, as already discussed [87].

In the absence of steady sources, the thermodynamical arrow of time is identified by the direction in which temperature gradients are smoothed out and eventually disappear. Macroscopic objects composed of many particles or subsystems are subject to this time orientation, even though the fundamental microscopic equations are time–reversible. It is curious that in the analogies between the Friedmann equation and the various phenomenological thermal models discussed, the thermodynamical arrow of time is analogous to the cosmological arrow of time (except that the real universe expands instead of contracting). In these fictitious universes, comoving objects are dragged by the cosmic contraction and the proper distances separating them eventually reduce to nothing as the scale factor $a(t)$ vanishes asymptotically at late times. This feature parallels the decaying temperature differences θ in models of convective–radiative heating and cooling.

The class of cooling models could be enlarged: we have discussed only models in which the function $f^2(\theta)$ appearing in Eq. (5.11) is a finite sum of power laws or inverse power laws, but more general $f(\theta)$ are possible. Then, imposing the energy conservation equation (1.187) would lead to non-linear barotropic equations of state $P = P(\rho)$ for the fluid filling the analogous universe. Non-linear equations of state have been studied in cosmology in relation with hypothetical forms of dark energy [26, 29, 36, 70, 185, 243, 299, 364, 398]. Equations of state of the form

$$P = \sum_{k=1}^{m} c_k \, \rho_{(k)}^k \qquad (3.65)$$

and their cosmological consequences are discussed in Refs. [55, 86, 87, 170, 386]. Energy conservation then requires that

$$\int \frac{d\rho}{\sum_{k=1}^{m} c_k \, \rho_{(k)}^k + \rho} = -3 \ln a \,, \qquad (3.66)$$

which is not easy to interpret. As special cases, quadratic equations of state are more transparent [6, 7, 79, 297, 298, 372], while pressures proportional to fractional powers of the energy density appeared in [299], but are even more speculative.

3.3   Cooling of lava flows

Lava flows are a very interesting phenomenon in volcanology. During volcanic eruptions, hot lava flows along the slopes of the volcano, often covering long distances, and cools. Over the course of many eruptions, lava flows contribute to building up the slope and give rise to a characteristic environment (Fig. 3.1).

Lava flows cool as they move and eventually form a crust at their surface, and possibly also lava tubes with hot lava flowing inside, insulated from their surroundings by this crust.

The dominant processes in lava cooling are radiation and convection into the atmosphere [200] which implies that, if the surface temperature of the lava is T, the heat flux is

$$\frac{dQ}{dt} = \epsilon \sigma \left(T^4 - T_s^4 \right) + h \left(T - T_s \right), \qquad (3.67)$$

where ϵ is the greybody factor (or emissivity), σ is the Stefan–Boltzmann constant, T_s is the temperature of the atmosphere and h is a convection coefficient, assumed to be constant. Dividing by the heat capacity C and the surface area A and rewriting this equation in terms of the temperature difference $\theta(t) \equiv T(t) - T_S$, as already done in the previous sections, we obtain

$$\frac{d\theta}{dt} = \frac{\epsilon \sigma}{AC} \left[\left(\theta + T_s \right)^4 - T_s^4 \right] + \frac{h}{AC} \left(T - T_s \right). \qquad (3.68)$$

FIG. 3.1. Mount Etna (3,357 m) in Sicily, Italy.

Dividing by θ and squaring yields the analogue of the Friedmann equation

$$\left(\frac{\dot{\theta}}{\theta}\right)^2 = \left(\alpha\,\theta^3 + \beta\,\theta^2 + \gamma\,\theta + \delta\right)^2$$

$$= \alpha^2\theta^6 + 2\,\alpha\,\beta\,\theta^5 + \left(\beta^2 + 2\,\alpha\,\gamma\right)\theta^4 + 2\left(\beta\,\gamma + \alpha\,\delta\right)\theta^3$$

$$+ \left(\gamma^2 + 2\,\beta\,\delta\right)\theta^2 + 2\gamma\,\delta\,\theta + \delta^2\,, \tag{3.69}$$

where

$$\alpha = \frac{\epsilon\,\sigma}{AC}\,, \tag{3.70}$$

$$\beta = \frac{4\,\epsilon\,\sigma\,T_s}{AC}\,, \tag{3.71}$$

$$\gamma = \frac{6\,\epsilon\,\sigma\,T_s^2}{AC}\,, \tag{3.72}$$

$$\delta = \frac{4\,\epsilon\,\sigma\,T_s^3}{AC} + \frac{h}{AC}\,. \tag{3.73}$$

All the coefficients on the right hand side of Eq. (3.69) are positive, which means that the cosmological constant $\Lambda = 3\delta^2$ and the energy densities of the six fluids sourcing the analogous universe (which is spatially flat) are positive. All these fluids are exotic phantom fluids. Again, it makes sense to drop the higher powers of the temperature difference θ and we reduce to the situations of the previous sections. Lava flows exhibit a second, and different, analogy with FLRW cosmology that is discussed in Chap. 8.

3.4 Freezing of lakes in winter

There is a relatively simple model describing the natural phenomenon of freezing of lakes and bodies of water in winter and, based on the experience gained in the previous sections of this chapter, it is not too surprising that a cosmological analogy exists also for this phenomenon [147].

FIG. 3.2. Malga Boazzo Lake in the Trentino region of Italy.

The freezing of bodies of water in the natural environment (Fig. 3.2) has been studied for a long time in both the technical [11, 235, 238, 248, 254, 257, 265, 280, 326, 385] and the pedagogical [10, 49, 377, 427] literature. Although a detailed realistic description of lake freezing in winter or in mountainous and polar regions is made difficult by the variability of atmospheric conditions, boundaries, and chemical impurities in the water [11,235,238,248,254,257,265,280,326,385], a simplified one–dimensional model is often quite reasonable. The assumptions of this model are made explicit in the pedagogical literature [427] and include:

- The lake covers a large area and the boundary effects at its margins can be neglected.
- The lake is isolated, *i.e.*, not connected to other lakes, rivers, or bodies of water.
- The freshwater in the lake is chemically pure.
- The lake is deep. In practice, this assumption means depths larger than about ten meters and that short periods of cold weather are considered.[1]
- The geothermal flux from the lake bottom and the daily solar radiation are negligible.

The model is described by the following variables and parameters [427]: the ice density ρ_{ice}, the thermal conductivity of ice λ_{ice}, the latent heat of fusion of water L_f, the ice thickness z (measured from the water surface downwards), the temperature T_1 of the liquid water, and the air temperature T_3. A key simplification consists of describing the convective and radiative heat losses from the ice to the atmosphere as a single heat flux density. The latter is assumed to be proportional to the difference between air and ice temperatures [427] and described by a single heat coefficient h. This linear approximation is the main simplification of the model and it makes possible an analytic discussion. In reality, both fluxes are non-linear and modelling convection realistically is difficult because of changes in the air conditions, wind, and so on. In calm conditions, however, the simplified model of [427] becomes more realistic.

Under the previous assumptions, equating the heat flux density that escapes the ice surface into the atmosphere (due to radiation and convection) with the flux density of the heat diffusing from the water to the air through the ice, produces the simple equation describing the time dependence of the ice thickness $z(t)$ [427]:

$$\frac{dz}{dt} = \frac{2\,(T_1 - T_3)}{\rho_{\text{ice}}\, L_f}\, \frac{1}{\frac{1}{h} + \frac{z}{\lambda_{\text{ice}}}}. \tag{3.74}$$

Let us write this equation in terms of the quantities

---

[1]This assumption can be relaxed, allowing one to consider lakes freezing to the bottom or permanently frozen, long periods of cold weather as in polar/subpolar climates or in alpine regions [427].

$$\alpha \equiv \frac{2\,(T_1 - T_3)}{\rho_{\text{ice}}\,L_f\,\lambda_{\text{ice}}}\,, \tag{3.75}$$

$$y \equiv \frac{z}{\lambda_{\text{ice}}}\,, \tag{3.76}$$

$$y_0 \equiv \frac{1}{h}\,, \tag{3.77}$$

$$s \equiv y + y_0 = \frac{z}{\lambda_{\text{ice}}} + y_0\,. \tag{3.78}$$

$$\tag{3.79}$$

By squaring, Eq. (3.74) assumes the form

$$\left(\frac{\dot{s}}{s}\right)^2 = \frac{\alpha^2}{s^4} \tag{3.80}$$

analogous to the Friedmann equation (1.179) for a $\kappa = 0$, radiation–dominated FLRW universe.

3.4.1   *The analogous radiation–dominated universe*

With the identification $\big(t, s(t)\big) \longrightarrow \big(t, a(t)\big)$, we have a formal analogy between the Friedmann equation and Eq. (3.80), but for this analogy to be valid also the energy conservation equation (1.187) must be satisfied, which imposes

$$\rho(t) = \frac{\rho_0}{a^4(t)}\,, \tag{3.81}$$

i.e., the cosmic fluid of the analogous universe must be blackbody radiation, the energy density and pressure of which scale as $1/a^4$ [82,83,244,260,435]). The constant ρ_0 is determined as

$$\rho_0 = \frac{3\,\alpha^2}{8\pi G} \tag{3.82}$$

(as follows from Eqs. (3.81) and (3.80)), making the energy density positive (this should not be taken for granted as shown, for example, by the formal analogies examined in the previous chapter for the logistic equation).

In this analogy, the FLRW universe is spatially flat and the quantity

$$s(t) = \frac{z(t)}{\lambda_{\text{ice}}} + \frac{1}{h} \tag{3.83}$$

is the analogue of the scale factor $a(t)$, while $\dot{s}/s$ is analogous to the Hubble function $H \equiv \dot{a}/a$.

The solution of Eq. (3.74) with the initial condition $z(0) = 0$ [427] corresponds to the textbook solution for the scale factor of a spatially flat, radiation–dominated universe [82, 244, 260, 435]

$$z(t) = \lambda_{\text{ice}} \left[\sqrt{\frac{2\alpha\, t}{\lambda_{\text{ice}}} + \frac{1}{h^2}} - \frac{1}{h} \right] , \qquad (3.84)$$

which can also be written as

$$s(t) = \sqrt{\frac{2\alpha\, t}{\lambda_{\text{ice}}} + \frac{1}{h^2}} \qquad (3.85)$$

with the initial condition

$$s(0) = \frac{1}{h} \equiv y_0 . \qquad (3.86)$$

In this model, freezing begins at $t = 0$ at the water surface $z = 0$, which corresponds to the finite value $a_0 = 1/h$ of the scale factor in the analogous universe, while the Big Bang $a = 0$ corresponds to $s = 0$ and

$$t_0 = -\frac{\lambda_{\text{ice}}}{2\alpha\, h^2} , \qquad (3.87)$$

$$z_0 = -\frac{\lambda_{\text{ice}}}{h} , \qquad (3.88)$$

$$y = -y_0 . \qquad (3.89)$$

Musing on this analogy, one can view the state of the universe as analogous to some "phase of gravity", with the liquid–solid phase transition of water possibly being analogous to some high energy phase transition. Indeed, phase transitions exist in both speculative and established high energy physics. For example, in string gas cosmology the early universe contains a gas of strings and a phase transition occurs at the critical Hagedorn temperature. In the high temperature Hagedorn phase, corrections to the low–energy effective action of string theory dominate but, as the temperature falls below the Hagedorn temperature and the universe cools after the phase transition, the fundamental string states become irrelevant, while brane states become important (see Refs. [33, 59] for reviews).

In another context, an extensive literature attempts to describe spacetime as a collective entity emerging from unspecified building blocks which can be likened to "atoms" or "molecules" of spacetime. In the causal set approach to quantum gravity [392], spacetime is discrete and made of isolated points which are not all in an order relation with each other. A causal set is a partially ordered system, a sort of "random lattice". It has been discovered [391] that, as the number N of points in a two–dimensional causal set increases, there is a phase transition in the system. In the limit of large N, the system acquires the properties of a continuum [391].

Another very intriguing and impactful idea is that of spacetime thermodynamics [130, 227, 313], in which it makes sense to speculate about phase transitions between different "phases of spacetime", in the same way that water has different phases. In the modest context of our formal analogy, the phase transition analogous to the one between liquid water and solid ice in freezing lakes begins at the value a_0 of the scale factor, after which the ice thickness z grows in real lakes and the analogous universe expands to $a > a_0$. A negative ice thickness $z < 0$ is meaningless and, similarly, the spacetime manifold view of the universe is not meaningful if $a < a_0$, which would correspond to gravity and spacetime being in a different phase. If such a vague picture of the universe could make sense in some physical implementation, the problem of the Big Bang singularity would be eliminated by the different "phase of spacetime" in the early universe for $a(t) < a_0$, to which the classical Einstein–Friedmann equations would not apply.

In the freezing lake analogy, the absence of ice and the persistence of freezing vaguely correspond to the idea of a non–classical static "universe" with $a = a_0$ at all early times and, therefore, to an asymptotic static past. It has been suggested in the literature that the Einstein static universe could be the initial state of the universe [30, 127]. However, the analogy breaks down with regard to heat fluxes: for freezing lakes, the phase change is accompanied by a heat flux from the cooling body of water, which is not isolated, to the atmosphere. By contrast, the universe is necessarily isolated, but it cools because of its expansion.

3.4.2 *Symmetries*

We can discuss in detail, for the radiation fluid, one of the symmetries presented for more general spatially flat FLRW cosmology. For a radiation fluid, this symmetry can be interpreted in relation with the conformal invariance of the massless photons composing it and has a partner in the freezing lake analogy.

Since the Maxwell equations ruling the electromagnetic field are conformally invariant in four spacetime dimensions [82, 93, 435], a radiation fluid is a conformally invariant form of matter. However, the Einstein equations ruling gravity are not conformally invariant. Nevertheless, a symmetry vaguely related to conformal invariance survives for a radiation fluid [147].

First, perform a conformal rescaling of the spatially flat FLRW metric $g_{\mu\nu} \to \tilde{g}_{\mu\nu} = \Omega^2 g_{\mu\nu}$ with conformal factor $\Omega(x^\alpha) > 0$. The new metric tensor $\tilde{g}_{\mu\nu}$ is not a solution of the Einstein equations with the same form of matter, but some residual conformal symmetry survives. Using conformal time η, the two FLRW line elements are mapped into each other,

$$ds^2 = a^2(\eta)\left(-d\eta^2 + d\vec{x}^2 \right) \to d\tilde{s}^2 = \Omega^2 a^2\left(-d\eta^2 + d\vec{x}^2 \right). \qquad (3.90)$$

In general, $d\tilde{s}^2$ is no longer a FLRW line element but, if the conformal factor Ω depends only on the time coordinate η, then the FLRW line element is mapped into another FLRW line element with new scale factor $\tilde{a}(\eta) = \Omega(\eta)\, a(\eta)$ because

$$d\tilde{s}^2 = \tilde{a}^2(\eta)\left(-d\eta^2 + d\vec{x}^2 \right). \qquad (3.91)$$

In general, $\tilde{a}(\eta)$ does not satisfy the Einstein–Friedmann equation with a radiation fluid. Further consider the special conformal factor $\Omega = a(\eta)$: in this case (using $s(t)$ as a synonymous of $a(t)$), the transformation

$$s \to \tilde{s} = s^2, \qquad s = \sqrt{\tilde{s}}, \qquad (3.92)$$

$$dt \to d\tilde{t} = s^2 dt, \qquad (3.93)$$

preserves the nature of the radiation fluid. To wit,

$$\frac{ds}{dt} = \frac{ds}{d\tilde{t}}\frac{d\tilde{t}}{dt} = \frac{\sqrt{\tilde{s}}}{2}\frac{d\tilde{s}}{d\tilde{t}} \qquad (3.94)$$

and Eq. (3.80) is mapped into

$$\frac{d\tilde{s}}{d\tilde{t}} = \frac{\tilde{\alpha}}{\tilde{s}} \qquad (3.95)$$

where $\tilde{\alpha} = 2\alpha$. The Friedmann equation with the new variables is

$$\left(\frac{1}{\tilde{s}}\frac{d\tilde{s}}{d\tilde{t}} \right)^2 = \frac{\tilde{\alpha}^2}{\tilde{s}^4} \qquad (3.96)$$

and it remains a Friedmann equation for a radiation fluid. Furthermore, the operation

$$\rho - \frac{\rho_0}{s^4} \to \tilde{\rho} = \frac{\rho}{\tilde{s}^4} = \frac{\tilde{\rho}}{s^8} \qquad (3.97)$$

leaves the energy conservation equation for a radiation fluid [138] unchanged as

$$\dot{\rho} + 4H\rho = 0 \qquad (3.98)$$

becomes

$$\frac{d\tilde{\rho}}{d\tilde{t}} + \frac{4}{\tilde{s}}\frac{d\tilde{s}}{d\tilde{t}}\tilde{\rho} = 0 \,. \tag{3.99}$$

It is well known that under conformal transformations of FLRW space $g_{\mu\nu} \to \tilde{g}_{\mu\nu} = \Omega^2 g_{\mu\nu}$, the energy density and pressure of the cosmic fluid transform as [133]

$$\tilde{\rho} = \Omega^{-4}\rho \,, \tag{3.100}$$

$$\tilde{P} = \Omega^{-4}P \,, \tag{3.101}$$

preserving the equation of state $P = w\,\rho$ of the cosmic fluid. In particular, if the fluid is radiation and $\Omega = a(\eta)$, it is

$$\rho \to \tilde{\rho} = \Omega^{-4}\rho = \Omega^{-4}\frac{\rho_0}{s^4} = \frac{\rho_0}{s^8} = \frac{\rho_0}{\tilde{s}^4} \,. \tag{3.102}$$

In summary, the variables change

$$s \to \tilde{s} = s^2 \,, \tag{3.103}$$

$$dt \to d\tilde{t} = s^2\,dt \,, \tag{3.104}$$

$$\rho \to \tilde{\rho} = \frac{\rho}{s^4} = \frac{\rho}{\tilde{s}^2} \,, \tag{3.105}$$

preserves the form of the Einstein–Friedmann equations for a spatially flat, radiation–dominated FLRW universe. Using the analogy between freezing lakes and FLRW cosmology, this symmetry transfers to the solutions of Eq. (3.74) [147]. Specifically, the solution

$$s(t) = \sqrt{\frac{2\alpha\,t}{\lambda_{\text{ice}}} + \frac{1}{h^2}} \tag{3.106}$$

of Eq. (3.80) is mapped by the symmetry operation into

$$\tilde{s}(t) = s^2(t) = \frac{2\alpha\,t}{\lambda_{\text{ice}}} + \frac{1}{h^2} \,. \tag{3.107}$$

We must now write $\tilde{s}$ in terms of the rescaled time as $\tilde{s}(\tilde{t}) = \tilde{s}\left(t(\tilde{t})\right)$. To this end, we integrate the relation $d\tilde{t}/dt = \tilde{s}$, which yields[2]

$$\tilde{t} = \int dt\,\tilde{s} = \int dt\,s^2(t) = \int dt\left(\frac{2\alpha\,t}{\lambda_{\text{ice}}} + \frac{1}{h^2}\right) = \frac{\alpha\,t^2}{\lambda_{\text{ice}}} + \frac{t}{h^2} \,. \tag{3.108}$$

---

[2]We choose the same origin $t = \tilde{t} = 0$ for t and $\tilde{t}$.

This is the quadratic equation

$$\frac{\alpha t^2}{\lambda_{\text{ice}}} + \frac{t}{h^2} - \tilde{t} = 0 \tag{3.109}$$

for t, with roots

$$t(\tilde{t}) = \frac{\lambda_{\text{ice}}}{2\alpha} \left(-\frac{1}{h^2} \pm \sqrt{\frac{4\alpha\tilde{t}}{\lambda_{\text{ice}}} + \frac{1}{h^4}} \right) . \tag{3.110}$$

Choosing the positive sign of the square root so that $\tilde{t} = 0$ corresponds to $t = 0$, we have

$$\tilde{s}(\tilde{t}) = \frac{2\alpha\,t(\tilde{t})}{\lambda_{\text{ice}}} + \frac{1}{h^2} = \sqrt{\frac{2\tilde{\alpha}\,\tilde{t}}{\lambda_{\text{ice}}} + \frac{1}{\tilde{h}^2}} , \tag{3.111}$$

where $\tilde{\alpha} = 2\alpha$ and $\tilde{h} = h^2$. When we translate these relations into equations for the ice thickness, we have that

$$z(t) = \lambda_{\text{ice}} \left[s(t) - \frac{1}{h} \right] = \lambda_{\text{ice}} \left[\sqrt{\frac{2\alpha t}{\lambda_{\text{ice}}} + \frac{1}{h^2}} - \frac{1}{h} \right] \tag{3.112}$$

is mapped to

$$\tilde{z}(\tilde{t}) = \lambda_{\text{ice}} \left[\sqrt{\frac{2\,\tilde{\alpha}\,\tilde{t}}{\lambda_{\text{ice}}} + \frac{1}{\tilde{h}^2}} - \frac{1}{\tilde{h}} \right] \tag{3.113}$$

by the symmetry operation (3.103)–(3.105), leaving the solution invariant in form.

Chapter 4

Cosmic analogies in gravity

*I especially love analogies, my
most faithful masters,
acquainted with all the secrets
of nature ... One should make
great use of them.*

Johannes Kepler

4.1 Falling raindrops, debris flows, and avalanches

The vertical velocity of a falling raindrop obeys an ordinary differential
equation analogous to a Friedmann equation [301]. This same equation
rules, under realistic approximations, rapid flow landslides, debris flow,
and avalanches.

4.1.1 *Falling raindrop*

Consider a raindrop of mass m falling vertically in air with velocity
$v(t) = dy/dt$, where $y(t)$ is its position at time t and the vertical y–axis
points downwards. After a short time, the friction force is proportional to
the square of the drop's velocity, $F_{\text{friction}} = -\epsilon\,v^2$. Neglecting the buoyant
force, which is irrelevant here, the equation of motion of the raindrop is

$$m\,\frac{d^2y}{dt^2} - mg - \epsilon\,v^2\,, \tag{4.1}$$

where g is the (constant) acceleration of gravity, or

$$\frac{dv}{dt} + \frac{\epsilon}{m}\,v^2 - g = 0\,, \tag{4.2}$$

which is a Riccati equation [210, 222]. We can rewrite it as

$$\left(\frac{\dot{v}}{v}\right)^2 = \frac{g^2}{v^2} + \frac{\epsilon^2}{m^2} v^2 - \frac{2\,g\,\epsilon}{m}, \tag{4.3}$$

which is analogous to the Friedmann equation [301]

$$H^2 = -\frac{K}{a^2} + \frac{8\pi G \rho_0}{3} a^2 + \frac{\Lambda}{3} \tag{4.4}$$

if we identify the drop velocity $v(t)$ with the scale factor $a(t)$ and we set

$$\Lambda = -\frac{6\,g\,\epsilon}{m}, \tag{4.5}$$

$$K = -g^2, \tag{4.6}$$

$$\frac{8\pi G \rho_0}{3} = \left(\frac{\epsilon}{m}\right)^2. \tag{4.7}$$

The analogous universe has hyperbolic spatial sections, negative cosmological constant, and contains a phantom fluid with positive energy density and equation of state parameter $w = -5/3$ (the acceleration equation (1.180) is automatically satisfied).

The falling raindrop reaches terminal speed (formally, only asymptotically but, in practice, in a short time)

$$v_{\mathrm{T}} = \sqrt{\frac{m\,g}{\epsilon}}, \tag{4.8}$$

which is obtained by setting $\dot{v}$ to zero in Eq. (4.2). The analogous universe has asymptotically linear scale factor $a_\infty(t) = \dot{a}_0\, t$ and $H_\infty = 0$, which is obtained by the three terms in the right hand side of Eq. (4.4) balancing out to add to zero.

The Riccati equation is solved by setting

$$v = \frac{m}{\epsilon}\frac{\dot{u}}{u}, \tag{4.9}$$

which reduces (4.2) to the equation $\ddot{u} - \frac{\epsilon\,g}{m}\,u = 0$ describing an inverted harmonic oscillator, with general solution

$$u(t) = A\,e^{t/\tau} + B\,e^{-t/\tau}, \tag{4.10}$$

where $\tau = \sqrt{\dfrac{m}{\epsilon\,g}}$ is a time scale and A, B are arbitrary integration constants. Correspondingly, the drop velocity and position are

$$v(t) = \sqrt{\frac{mg}{\epsilon}} \left(\frac{A\,e^{t/\tau} - B\,e^{-t/\tau}}{A\,e^{t/\tau} + B\,e^{-t/\tau}} \right) \tag{4.11}$$

and

$$y(t) = y_0 + \frac{m}{\epsilon} \ln\left(A\,e^{t/\tau} + B\,e^{-t/\tau} \right), \tag{4.12}$$

where y_0 is the initial position. As $v(t) \to v_T$ when $t \to +\infty$, the scale factor of the corresponding universe becomes linear, $a(t) \simeq \dot{a}_0\, t$.

If the raindrop begins its fall from rest at y_0, then $A = B = 1/2$,

$$v(t) = g\,\tau \tanh\left(\frac{t}{\tau} \right) \tag{4.13}$$

and

$$y(t) = y_0 + \frac{m}{\epsilon} \ln\left[\cosh\left(\frac{t}{\tau} \right) \right]. \tag{4.14}$$

The corresponding scale factor can be added to the catalog of multi–fluid solutions of the Einstein–Friedmann equations [157] for the exotic phantom fluid with $w = -5/3$, $\Lambda < 0$, and negative curvature.

The analogy works also for a charged particle in constant vertical electric field (for which it is possible to switch the sign of the vertical forces). It is also possible to generate another analogy with the Friedmann equation written in conformal time η. In general, the Friedmann equation for a single fluid with constant equation of state, curvature, and $\Lambda = 0$, when rewritten in terms of η instead of comoving time t assumes the form of a Riccati equation, which offers an alternative way to its solution [132].

The falling raindrop forms a paradigm for any particle that is subject to a constant force and to a viscous drag quadratic in the velocity. For example, one can think of a vehicle moving on a horizontal surface at relatively high speed (so that the air drag is quadratic in the velocity) and propelled by a constant force (this would be the case if the motor was working at constant power output, for example at its maximum output). Another scenario with the same description is that of a boat speeding on flat water at maximum power output, which quickly reaches a terminal speed. Indeed natural systems often adjust to a terminal speed, as discussed in the next subsection.

4.1.2   *Landslides, debris flows, and avalanches*

In rapid flow landslides, debris flows, or certain avalanches, the velocity of the flow is extremely important because it controls the dynamics of the landslide and it is a crucial quantity to measure in order to assess, and then mitigate, the hazard posed by a slide. This is possible, for example, on slopes that avalanche regularly.

Landslides obey partial differential equations but there is a centre of mass description in which the sliding mass is treated as homogeneous, *i.e.*, effectively, as a point particle [329, 426]. This approximate description captures features present in most landslide models that are, in this sense, robust [339].

Consider a slide along an inclined plane extending infinitely far in the transversal direction and let x be a longitudinal coordinate pointing downward along the flow. Let $u(t, x)$ be the velocity of the material, modelled as a fluid, and assume viscous drag proportional to the square of the velocity. It can be shown [329, 339, 426] that the velocity obeys the equation (obtained by combining the continuity equation expressing mass conservation and the momentum balance equation)

$$\frac{\partial u}{\partial t} + u\frac{\partial u}{\partial x} = \alpha - \beta\, u^2\,, \qquad (4.15)$$

where α and β are constants. If there is negligible local deformation, $\partial u/\partial x \simeq 0$, the point mass approximation is valid, and the equation reduces to [329, 339, 426]

$$\dot{u} = \alpha - \beta\, u^2\,, \qquad (4.16)$$

which is the equation of motion for a falling drop already discussed. Although the contexts are different, the analogy between a falling raindrop and the slide is almost perfect since in both cases gravity (or a component of it along the incline) causes the motion and the friction is, in both cases, quadratic in the velocity.

It is realistic to think that, if the slide starts with zero initial velocity, then in the initial instants in which u is small the drag force proportional to u^2 can be neglected (as for the raindrop that is essentially in free fall before its speed increases, along with the friction force acting on it). Then the slide in the centre of mass approximation is equivalent to a particle in free fall with acceleration equal to the component of the acceleration of gravity along the slope, and the velocity as a function of the distance travelled is [339]

$$u(x) = \sqrt{2\alpha\,(x - x_0)}\,, \qquad (4.17)$$

a formula familiar from the motion of falling bodies in first year physics courses. As the slide picks up speed, the viscous drag quickly becomes quadratic in u and slows down the slide to the terminal velocity $u_T = \sqrt{\alpha/\beta}$. It may be surprising that this happens but, if the slope is sufficiently long and the slide lasts for a sufficiently long time, it is not uncommon to see a mass of debris, mud, or snow sliding down a slope at constant speed.[1] Since the terminal speed is approached from below, it is an upper limit on the velocity of the slide and it is often used to estimate the maximum velocity of an avalanche [278, 338, 426], which simplifies considerably the dynamical problem and facilitates the assessment of hazards. The centre of mass approximation is valid until the slide fragments or deforms substantially.

Since the sliding velocity equation is the same as that of a falling raindrop, the analogy with a hyperbolic FLRW universe with negative cosmological constant and dominated by a phantom fluid still applies (remember, however, that the analogue of the scale factor is not the position of the slide, but its velocity).

4.2 Brachistochrone problem

A textbook problem of variational calculus consists of determining the curve in the vertical plane that connects two given points (not on the same vertical), and such that a point particle sliding on this curve without friction reaches the bottom in the minimum time. Such a curve is called a brachistochrone [42, 183]. This problem was the subject of a contest organized by Johann Bernoulli for the elite mathematical community in 1696 [349].

Let v be the speed of a particle of mass m falling freely from rest, *i.e.*, $v_0 = 0$, from the height y (not following the brachistochrone). The conservation of energy gives $m\,v^2/2 = m\,g\,y$, where g is the constant acceleration of gravity, and then

$$v = \sqrt{2\,g\,y}. \tag{4.18}$$

We want to extremize the descent time along a curve $y(x)$ in the vertical plane from point 1 to point 2; the infinitesimal arc length along this curve is $d\ell = \sqrt{dx^2 + dy^2}$ and the transit time is obtained by integrating along

---

[1]The author remembers a flow of fairly consolidated snow sliding down a very long mountain slope in a regular way in a stream that lasted a relatively long time. As observed from a vantage point quite low on the slope, the speed was constant and the flow was very regular.

the curve from point 1 to point 2,

$$J\left[y(x)\right] = \int_1^2 dt = \int_1^2 \frac{d\ell}{v} = \int_{x_1}^{x_2} dx \sqrt{\frac{1 + y'^2}{2\,g\,y}}$$

$$\equiv \int_{x_1}^{x_2} dx\, L\left(y(x), y'(x)\right), \tag{4.19}$$

where a prime denotes differentiation with respect to x and $y > 0$. The descent time is extremized by imposing its variation to vanish along the extremizing path, $\delta J = 0$, which is equivalent to imposing that the Lagrangian L satisfies the Euler–Lagrange equation. We do not need to solve the latter because this Lagrangian does not depend explicitly on the coordinate x, which leads to conservation of the corresponding Hamiltonian. This first integral of motion (Beltrami identity) reads

$$\mathcal{H} = \frac{\partial L}{\partial y'}\, y' - L = C_0\,, \tag{4.20}$$

where C_0 is a constant. It can be written as

$$\sqrt{y\left(1 + y'^2\right)} = C_1\,, \tag{4.21}$$

where now

$$C_1 = -\frac{1}{C_0\,\sqrt{2\,g}} > 0\,, \tag{4.22}$$

which yields the first order ordinary differential equation

$$y'^2 = \frac{C_1^2}{y} - 1\,. \tag{4.23}$$

Its solution is a cycloid, which is the prototypical roulette [183]. The reader has certainly spotted the way to obtain a cosmic analogy: we divide both sides of (4.23) by y^2, obtaining the Friedmann–like equation

$$\left(\frac{y'}{y}\right)^2 = \frac{C_1^2}{y^3} - \frac{1}{y^2} \tag{4.24}$$

analogous to

$$H^2 = \frac{8\pi G\rho_0}{3a^3} - \frac{1}{a^2}\,. \tag{4.25}$$

This Friedmann equation rules a $\kappa = +1$ FLRW universe filled with a dust with zero pressure and energy density $\rho = \rho_0/a^3$. Fortunately, ρ_0 and ρ are positive, or the physical analogy would be spoiled. The classic scale factor solution for this universe is given in parametric form with the conformal time η as the parameter,

$$a(\eta) = \frac{C_2}{2}\left(1 - \cos\eta\right), \tag{4.26}$$

$$t(\eta) = \frac{C_2}{2}\left(\eta - \sin\eta\right), \tag{4.27}$$

where $C_2 = 8\pi G\rho_0$ and we assumed the initial condition $a(t=0) = 0$.

One can interpret the parameter η analogous to the conformal time for the actual cycloid solution of the original brachistochrone problem as follows. η is defined by $dx = y\,d\eta$, meaning that small increments of η equal small increments of the x–coordinate measured in units of the height of the point $\left(x, y(x)\right)$ on the brachistochrone.

According to Eq. (4.23), $y' \to \infty$ as $y \to 0$, meaning that the brachistochrone has a cusp at its highest point. This feature has a parallel in the analogous dust–dominated universe, which must begin at a Big Bang singularity. This feature reflects the much more general Hawking–Penrose singularity theorems, according to which matter satisfying the null energy condition (including dust) cannot avoid such a singularity [435].

4.3   Gravity tunnels

Another variational problem comes from the pedagogical physics literature. It consists of finding the motion of a point particle through a tunnel dug out along a diameter of the Earth. For simplicity, the latter is treated as a uniform sphere [92, 239, 253, 270, 351, 415], but the case of a non–uniform sphere has also been discussed [241].

A homogeneous sphere of mass M and radius R has uniform density

$$\rho_s = \frac{3M}{4\pi R^3}; \tag{4.28}$$

the gravitational potential Φ corresponding to this mass distribution is spherically symmetric and satisfies the Poisson equation

$$\frac{1}{r^2}\frac{d}{dr}\left(r^2\frac{d\Phi}{dr}\right) = 4\pi G\rho_s \tag{4.29}$$

inside the sphere. The solution is easily found to be

$$\Phi(r) = \frac{GM}{2R^3} r^2 - \frac{C_1}{r} + C_2 \,, \tag{4.30}$$

with $C_{1,2}$ integration constants. The boundary condition that this potential is regular at the centre of the sphere imposes $C_1 = 0$. The potential outside the sphere is $\Phi_{\text{out}} = -GM/r$ and, matching it with the interior potential at the surface $r = R$ determines the second integration constant as $C_2 = -\frac{3GM}{2R}$. Then, the gravitational potential inside the homogeneous sphere is

$$\Phi(r) = \frac{GM}{2R^3} r^2 - \frac{3GM}{2R} \,, \tag{4.31}$$

which is recognized as a shifted harmonic oscillator potential. This conclusion implies that the point particle oscillates up and down the tunnel.

The Lagrangian for a particle of mass m and position $r(t)$ moving without friction through this gravity tunnel is simply the difference of its kinetic and potential energies [183]

$$L\left(r, \dot{r}\right) = \frac{m \, \dot{r}^2}{2} - \frac{GM}{2R^3} r^2 + \frac{3GM}{2R} \,. \tag{4.32}$$

This Lagrangian does not depend explicitly on time and, therefore, the corresponding Hamiltonian is the particle's total mechanical energy and is conserved,

$$\frac{\partial L}{\partial \dot{r}} \dot{r} - L = E \,. \tag{4.33}$$

Written explicitly, this energy integral of motion (or Beltrami identity) reads

$$\frac{m \, \dot{r}^2}{2} + m \left(\frac{GM}{2R^3} r^2 - \frac{3GM}{2R} \right) = E \tag{4.34}$$

and, as we are now accustomed, can be rewritten as

$$\left(\frac{\dot{r}}{r} \right)^2 = -\frac{GM}{R^3} - \frac{K}{r^2} \,, \tag{4.35}$$

where

$$K = -\frac{2}{m} \left(E + \frac{3GM}{R} \right) \,, \tag{4.36}$$

originating another cosmological analogy [145]. The cosmos satisfying the analogous Friedmann equation (1.179) has scale factor $a(t)$ analogous to $r(t)$ and obeying

$$H^2 = \frac{\Lambda}{3} - \frac{K}{a^2} \,. \tag{4.37}$$

This equation involves the negative cosmological constant

$$\Lambda = -\frac{3GM}{R^3} \tag{4.38}$$

and no matter. It is $H^2 \geq 0$, which implies that only a negative curvature index $K < 0$ can appear because the curvature term $-K/a^2$ must compensate the negative cosmological constant in the Friedmann equation. This analogous universe is the Anti–de Sitter spacetime made famous by the holographic principle and the AdS/CFT correspondence [220]. The explicit scale factor of the analogous universe which solves Eq. (4.37) is

$$a(t) = \sqrt{\frac{3}{|\Lambda|}} \, \sin\left(\sqrt{\frac{|\Lambda|}{3}} \, t\right). \tag{4.39}$$

4.4   Terrestrial brachistochrone

A more sophisticated version of the gravity tunnel variational problem, called the terrestrial brachistochrone problem [92, 415], combines the gravity tunnel with a variation of the brachistochrone problem discussed above. Tunnels of various curved shape, described by functions $r(\vartheta)$ in polar coordinates, are dug out in a uniform Earth of radius R; the variational problem consists of finding the tunnel with shape $r(\vartheta)$ minimizing the transit time between two given points. This problem has mostly pedagogical interest [8, 91, 104, 105, 109, 114, 180, 224, 225, 242, 362, 363, 373, 399] but, curiously, it has received some attention from the mining industry [161, 196, 447].

The Lagrangian for the terrestrial brachistochrone problem is found in Ref. [415],

$$L(r, r') = \sqrt{\frac{r^2 + r'^2}{R^2 - r^2}}, \tag{4.40}$$

where a prime denotes differentiation with respect to the polar coordinate ϑ. Given that $\partial L/\partial \vartheta = 0$, the Hamiltonian coincides with the particle energy and is conserved [183],

$$\mathcal{H} = \frac{\partial L}{\partial r'} r' - L = C. \tag{4.41}$$

Writing out this equation explicitly gives

$$\frac{r^2}{\sqrt{R^2 - r^2} \sqrt{r^2 + r'^2}} = -C, \tag{4.42}$$

which can be recast as the Friedmann–like equation

$$\left(\frac{r'}{r}\right)^2 = -1 + \frac{r^2}{C^2 (R^2 - r^2)}. \tag{4.43}$$

The travelling particle attains its minimum radius where the curve $r(\vartheta)$ flattens at $r' = 0$, which gives

$$C^2 = \frac{r_{\min}^2}{R^2 - r_{\min}^2}. \tag{4.44}$$

We then have

$$\frac{r^2 + r'^2}{r^2} = \frac{r^2}{r_{\min}^2} \frac{\left(R^2 - r_{\min}^2\right)}{\left(R^2 - r^2\right)}. \tag{4.45}$$

The Friedmann equation analogous to Eq. (4.43) is

$$H^2 = -1 + \frac{a^2}{C^2\left(a_0^2 - a^2\right)} \tag{4.46}$$

and describes a cosmos with negative cosmological constant $\Lambda = -3$ filled by a perfect fluid with energy density

$$\rho = \rho_0 \frac{a^2}{a_0^2 - a^2}, \tag{4.47}$$

where

$$\frac{8\pi G}{3} \rho_0 = \frac{1}{C^2}. \tag{4.48}$$

The cosmological analogy is not complete unless the covariant conservation equation (1.187) is satisfied, which determines the equation of state of the cosmic fluid in the analogous universe. Imposing covariant energy conservation yields

$$P = -\rho - \frac{2\rho_0 a_0^2 a^2}{3\left(a_0^2 - a^2\right)^2}. \tag{4.49}$$

We then use the relation

$$\frac{a^2}{\left(a_0^2 - a^2\right)^2} = \left(\frac{\rho}{\rho_0}\right)^2 \frac{1}{a^2} \tag{4.50}$$

and

$$a^2 = \frac{\rho}{\rho_0} \frac{a_0^2}{\left(1 + \rho/\rho_0\right)}, \tag{4.51}$$

which determines the phantom quadratic equation of state

$$P(\rho) = -\frac{5}{3}\rho - \frac{2}{3}\frac{\rho^2}{\rho_0}, \tag{4.52}$$

falling into the category studied in Refs. [6, 7, 78, 86, 87, 297, 298, 372, 398]. These non-linear equations of state are associated with specific types of cosmological singularities different from the more familiar Big Bang, Big Crunch, and Big Rip [16, 26, 29, 36, 56, 98, 99, 163, 356, 398].

The Ricci scalar of our analogous universe is

$$\mathcal{R} = -\rho_{\text{total}} + 3P_{\text{total}} = \frac{1}{2\pi G} - 6\rho - \frac{2\rho^2}{\rho_0} \tag{4.53}$$

and it diverges as $a \to a_0$. This is a clear signal of a spacetime singularity in which the scale factor stays finite but the Hubble function H, the energy density ρ, and the pressure P are infinite. This peculiar cosmological singularity has a counterpart in the terrestrial brachistochrone, where a_0 corresponds to the Earth radius R. In the analogous universe, the value a_0 of the scale factor cannot be crossed dynamically and $a(t) < a_0$ for all times t, but $a(t) \to a_0$ (see the discussion below). The scale factor $a(t)$ has a cusp at a_0, corresponding to the singularity in the right hand side of the Friedmann equation (4.46). Correspondingly, the terrestrial brachistochrone starts at the Earth surface $r = R$ (corresponding to $a = a_0$) normal to it and the derivative $r' \equiv dr/d\vartheta$ diverges there [241, 415].

Equation (4.46) tells us that there is also a minimum value for the scale factor $a(t)$ of the analogous cosmos: since $H^2 \geq 0$, values of the scale factor $a(t) < a_{\min}$ are forbidden, where

$$a_{\min}^2 = \frac{C^2 a_0^2}{1 + C^2} \tag{4.54}$$

is the value of the scale factor corresponding to $H = 0$. The dynamics of $a(t)$ is restricted to the strip

$$a_{\min} \leq a(t) < a_0 \tag{4.55}$$

of the $\left(t, a\right)$ plane. The constant C satisfies

$$C^2 = \frac{3}{8\pi G \rho_0} = \frac{a_{\min}^2}{a_0^2 - a_{\min}^2}. \tag{4.56}$$

On the terrestrial brachistochrone side of the analogy, the minimum value of $a(t)$ corresponds to the minimum radius $r_{\min}$ of the moving particle[2] [241, 415].

Interestingly, the static universe with constant scale factor $a(t) \equiv a_{\min}$ is also an analytical solution of the Einstein–Friedmann equations (1.179)–(1.187). It is a fine-tuned cosmology achieved by balancing the attraction of the cosmological constant $\Lambda < 0$ with the repulsion of exotic matter

[2]The terrestrial brachistochrone does not pass through the centre of the Earth $r = 0$ [241, 415].

$(\rho + 3P) < 0$ in the acceleration equation (1.180), in which these terms appear with opposite signs and compete. This equilibrium is achieved at the price of fine–tuning parameters and initial conditions and should, therefore, be unstable. This expectation is confirmed by a linear stability analysis, in which one finds an unstable exponentially growing mode (see Appendix B.1). The static solution $a(t) \equiv a_{\min}$ is the only equilibrium point of the dynamical system (1.179)–(1.187) under the assumptions above and is a repellor in the phase space of the solutions.

Continuing our analysis we note that, according to the acceleration equation (1.180), the analogous cosmos is accelerated, $\ddot{a} > 0$: since $a_{\min} < a(t) < a_0$, the function $a(t)$ is concave upward, $\dot{a}$ increases becoming larger and larger the closer the orbit of the solution comes to the singularity a_0. The orbits of solutions starting in the region $a_{\min} < a < a_0$ approach the singularity faster and faster (*i.e.*, with larger and larger $\dot{a}$), until they hit it in a finite proper time: here the universe meets its end, although the scale factor remains finite at the value a_0.

The cosmological side of the analogy, with its exotic universe filled by the phantom fluid with quadratic equation of state, benefits from the fact that the analytical solution of the terrestrial brachistochrone is known [145]. Its graph is a hypocycloid curve [241,415], which is the curve generated by a circle of radius $(R - r_{\min})/2$ rolling without sliding at constant speed inside the larger circle of radius R (another roulette, according to the established fact that all solutions of the Friedmann equation are roulettes [87]). Its parametric representation is [241,415]

$$r^2(t) = \frac{\left(R^2 + r_{\min}^2\right)}{2} - \frac{\left(R^2 - r_{\min}^2\right)}{2} \cos\left(2\,\omega\,t\right) , \tag{4.57}$$

$$\vartheta(t) = \tan^{-1}\left[\frac{R}{r_{\min}}\tan(\omega\,t)\right] - \frac{r_{\min}}{R}\,\omega\,t . \tag{4.58}$$

The parameter is the time t, $\omega = 2\pi/T$, and

$$T = \pi\sqrt{\frac{R^2 - r_{\min}^2}{g\,R}} \tag{4.59}$$

is the travel time between two points on the surface of the Earth, while g is the acceleration of gravity at the surface. T is the minimum travel time among all tunnel configurations between these two points on the surface [241,415].

Another parametric representation of the analytical solution using the radial coordinate r as the parameter is [241, 415]

$$\vartheta(r) = \tan^{-1}\left(\frac{R}{r_{\min}}\sqrt{\frac{r^2 - r_{\min}^2}{R^2 - r^2}}\right) - \frac{r_{\min}}{R}\tan^{-1}\left(\sqrt{\frac{r^2 - r_{\min}^2}{R^2 - r^2}}\right),$$

(4.60)

$$t(r) = \frac{\sqrt{R^2 - r_{\min}^2}}{2\sqrt{Rg}}\cos^{-1}\left(\frac{R^2 + r_{\min}^2 - 2r^2}{R^2 - r_{\min}^2}\right).$$

(4.61)

Using the cosmological analogy, the analytical solution for the scale factor of the analogous cosmos is

$$a^2(s) = \frac{(a_0^2 + a_{\min}^2)}{2} - \frac{(a_0^2 - a_{\min}^2)}{2}\cos(2\,\omega\,s),$$

(4.62)

$$t(s) = \tan^{-1}\left[\frac{a_0}{a_{\min}}\tan(\omega s)\right] - \frac{a_{\min}}{a_0}\,\omega\,s,$$

(4.63)

or, alternatively,

$$t(a) = \tan^{-1}\left(\frac{a_0}{a_{\min}}\sqrt{\frac{a^2 - a_{\min}^2}{a_0^2 - a^2}}\right) - \frac{a_{\min}}{a_0}\tan^{-1}\left(\sqrt{\frac{a^2 - a_{\min}^2}{a_0^2 - a^2}}\right),$$

(4.64)

$$s(a) = \frac{\sqrt{a_0^2 - a_{\min}^2}}{2\sqrt{a_0\,g}}\cos^{-1}\left(\frac{a_0^2 + a_{\min}^2 - 2a^2}{a_0^2 - a_{\min}^2}\right).$$

(4.65)

As the universe approaches the singularity $a = a_0$ it ends its existence in a finite time, as shown by the fact that the limit $a \to a_0$ of the parametric solution (4.62), (4.63) corresponds to the finite parameter value

$$s_0 = \pi\sqrt{\frac{a_0^2 - a_{\min}^2}{a_0\,g}}$$

(4.66)

and to the finite time

$$t_0 = \frac{\pi}{2}\left(1 - \frac{a_{\min}}{a_0}\right),$$

(4.67)

Setting $t_0 = T/2$ yields

$$g = a_0\left(\frac{a_0 + a_{\min}}{a_0 - a_{\min}}\right).$$

(4.68)

4.5   Newtonian analogy for FLRW cosmology

A heuristic introduction to cosmology is often presented in textbooks or in non–technical discussions in the form of a Newtonian analogy with FLRW cosmology (*e.g.*, [120, 207, 260, 288, 354]). Its essential ingredients are a sphere of uniform density expanding in empty space and the conservation of energy for a test particle on the surface of this sphere in Newtonian mechanics. The Newtonian energy conservation equation is analogous to the Friedmann equation (1.179) of relativistic cosmology for a FLRW universe filled with dust. We stress that this is only an analogy and has several limitations, above all the fact that the only cosmic fluid that can be discussed in this analogy is a dust, while the matter content of the universe can be quite different. Most, but not all, textbooks and lecture notes available on the internet stress the fact that this is only an analogy.

Let us consider, in Newtonian mechanics, a sphere with homogeneous density $\rho(t)$, radius $R(t)$, and total mass $M = 4\pi R^3 \rho/3$. Imagine a test particle of mass m on the surface of this sphere; its mechanical energy

$$E = \frac{1}{2} m \dot{R}^2 - \frac{GMm}{R} \tag{4.69}$$

is constant since gravity is a conservative force. This energy integral can be recast in the Friedmann–like form

$$\frac{\dot{R}^2}{R^2} = \frac{8\pi G}{3} \rho + \frac{2E}{mR^2} \,. \tag{4.70}$$

The analogy with the Friedmann equation is obtained by defining the quantities

$$H \equiv \frac{\dot{R}}{R}, \tag{4.71}$$

$$K \equiv -\frac{2E}{mc^2}, \tag{4.72}$$

where c is the speed of light, in terms of which the energy equation (4.70) is rewritten as

$$H^2 = \frac{8\pi G}{3} \rho - \frac{Kc^2}{R^2}, \tag{4.73}$$

which is formally the Friedmann equation for a dust–dominated universe. The possible signs of the energy E correspond to the three possibilities for the curvature of the FLRW spatial sections. Let

$$v_{\text{escape}} = \sqrt{\frac{2GM}{R}} \tag{4.74}$$

be the escape velocity from the surface of the ball. Then,

- The positive sign $E > 0$ of the energy corresponds to $K < 0$ and to $v > v_{\text{escape}}$: the particle escapes to infinite R with residual velocity $\dot{R}_\infty$ given by the limit

$$0 < E = \frac{1}{2} m \dot{R}^2 - \frac{GMm}{R} \to \frac{1}{2} m \dot{R}_\infty^2 . \tag{4.75}$$

- If, instead, $E = 0$, which corresponds to $K = 0$, flat FLRW spatial sections, and $v = v_{\text{escape}}$, the particle barely escapes to infinity with zero velocity $\dot{R}$, according to

$$0 = \frac{1}{2} m \dot{R}^2 - \frac{GMm}{R} \to 0 . \tag{4.76}$$

- Finally, if the particle energy is negative, $E < 0$, which corresponds to $K > 0$ and $v < v_{\text{escape}}$, then the particle reaches a maximum height and falls back, reversing its radial velocity. Accordingly, we cannot take the limit $R \to +\infty$ in this situation. The maximum radius is attained at the turning point $\dot{R} = 0$ given by

$$E = -\frac{GMm}{R_{\text{max}}} \tag{4.77}$$

and is

$$R_{\text{max}} = \frac{GMm}{|E|} . \tag{4.78}$$

Any cosmology textbook reports the analytical solutions of the energy integral (4.72) for these dust–dominated universes, in parametric form using conformal time η as the parameter. These solutions correspond to the possible radial motions of the test particle in the three situations described. Defining η by

$$d\eta = \sqrt{\frac{2|E|}{m}} \frac{dt}{R(t)} , \tag{4.79}$$

the solutions of the Newtonian problem are

$$R = \frac{GMm}{2|E|} \left(1 - \cos\eta \right) , \tag{4.80}$$

$$\pm (t - t_0) \sqrt{\frac{2|E|}{m}} = \frac{GMm}{2|E|} \left(\eta - \sin\eta \right) \tag{4.81}$$

if $E < 0$;

$$R(t) = R_0 \left(t - t_0 \right)^{2/3} \qquad (4.82)$$

(with R_0 a constant) if $E = 0$; and

$$R = \frac{GMm}{2\,|E|} \left(\cosh \eta - 1 \right), \qquad (4.83)$$

$$\pm\,(t - t_0)\,\sqrt{\frac{2E}{m}} = \frac{GMm}{2E} \left(\sinh \eta - \eta \right), \qquad (4.84)$$

if $E > 0$.

The energy E is a constant of motion of the Newtonian problem and the three possible radial motions are, therefore, mutually exclusive. This simple feature parallels the fact that the topology of the FLRW spatial sections cannot change. It is not dynamical and, as already remarked, it is impossible for a closed universe to become open and *vice–versa*, or for a closed or open universe to become flat, and *vice–versa*.

The Newtonian solution for $E < 0$ corresponds to an open, dust–dominated FLRW universe that expands forever; $E > 0$ mirrors a universe that reaches a maximum size and collapses in a Big Crunch, while $E = 0$ mocks a spatially flat FLRW universe. The Newtonian analogy provides some intuition but should not be extrapolated: the physical meaning of variables with the same name is different in the two sides of the analogy and the two contexts are completely separated. A FLRW universe has no centre of expansion, contrary to the Newtonian ball. This analogy, however, does provide a certain degree of physical insight and appears frequently in the literature. It is even used as a toy model in quantum cosmology [419, 420].

4.6   Turning Newtonian cosmology into a relativistic problem

After learning general–relativistic cosmology, one can ponder whether the Newtonian analogy for dust–dominated FLRW universes can be extended to a full–fledged general–relativistic problem. Toying with this idea involves considering a relativistic isolated ball expanding in empty space [152]. Or, one is led to a simpler problem if one remembers that the Newtonian analogy studies the motion of a test particle in the spherically symmetric field of the ball, which leads to studying geodesics instead [152]. Let us discuss these possibilities, beginning with the simplest scenario.

4.6.1   *Test particle in radial motion above a ball*

The simplest way to promote the Newtonian problem to a general–relativistic one consists of considering radial geodesics. In fact, the Newtonian analogy proceeds if the test particle is just above the surface of the ball instead of being exactly on it, and it moves radially away from this surface. As long as the expanding ball does not reach this particle and absorb it, or a falling particle that failed its escape hits the ball, the Newtonian problem makes sense (it is not granted that this does not happen, as will be discussed below). We assume that the test particle and the ball do not meet, at least for some time.

In the general–relativistic context, we consider a test particle in radial motion outside the ball, which starts out close to its surface and outgoing. If the sphere contracts, it does not meet the particle at least until the latter attains its maximum height and falls back (which may never happen). If this particle does fall back down, it meets the contracting sphere only if it falls radially faster than this sphere contracts (this may be possible if the material composing the ball has pressure and is non–geodesic). Other possibilities include a falling particle hitting an expanding ball, or the surface of the sphere expanding faster than the massive test particle moves. Separating the test particle from the surface of the sphere opens up these new possibilities. Let us assume, for now, that the test particle and the surface of the sphere do not collide.

Due to the Birkhoff theorem [120,435], the geometry of the empty spacetime outside the sphere is necessarily the Schwarzschild one with line element

$$ds^2 = -\left(1 - \frac{2m}{\bar{r}}\right)dt^2 + \frac{d\bar{r}^2}{1 - 2m/\bar{r}} + \bar{r}^2 d\Omega_{(2)}^2 \,, \qquad (4.85)$$

where m is the mass of the sphere. A massive test particle moves along a timelike geodesic of equation [435]

$$\dot{\bar{r}}^2 + \left(1 - \frac{2m}{\bar{r}}\right)\left(\frac{L^2}{\bar{r}^2} + \kappa\right) = E^2 \,. \qquad (4.86)$$

For completeness, we will discuss also radial null geodesics. Here an overdot denotes differentiation with respect to the proper time τ or the affine parameter along the geodesic, $\kappa = 1$ for timelike geodesics and $\kappa = 0$ for null geodesics, while E and L are the conserved energy and angular momentum per unit mass of the particle [82, 120, 435]. Apart from the sign, they arise from contracting the particle four–velocity u^μ with the timelike and rotational Killing vector fields of Schwarzschild spacetime, respectively [435].

4.6.1.1 *Massive test particle*

Timelike radial geodesics satisfy the equation

$$\left(\frac{\dot{\bar{r}}}{\bar{r}}\right)^2 = \frac{E^2 - 1}{\bar{r}^2} + \frac{2m}{\bar{r}^3}, \tag{4.87}$$

which is formally a Friedmann equation (1.179)

$$H^2 = \frac{8\pi\rho_0}{3\,a^3} - \frac{K}{a^2} \tag{4.88}$$

for a FLRW universe with curvature index $K = 1 - E^2$, energy density $\rho = \rho_0/a^3$ of a dust fluid, and $\rho_0 = \dfrac{3m}{4\pi}$. As in the Newtonian context, an analogy exists between the motion of a test particle in the field of the ball and the Friedmann equation. Again, only a dust–dominated analogous universe can be obtained, and all possible values of the curvature index K of the analogous FLRW universe are possible.

4.6.1.2 *Massless test particle*

Let us consider also radial null geodesics, which technically do not belong to Newtonian physics. An expanding fluid sphere will never reach photons starting radially above it, which always escape to infinity. The trajectories of these photons obey[3]

$$\dot{\bar{r}}^{\,2} = E^2. \tag{4.89}$$

Clearly, the cosmological analogue of this radial null trajectory is given by the Friedmann equation

$$H^2 = -\frac{K}{a^2} \tag{4.90}$$

where, necessarily, K is negative. This analogous cosmos is the Milne universe, which is nothing but the empty Minkowski spacetime seen by accelerated observers and sliced using a hyperbolic foliation [288]. The corresponding scale factor is linear, $a(t) = a_0\,t$, and the Riemann tensor vanishes identically. The corresponding FLRW line element in hyperspherical coordinates is

$$ds^2 = -dt^2 + t^2\left(d\chi^2 + \sinh^2\chi\,d\Omega_{(2)}^2\right), \tag{4.91}$$

which is reduced to the Minkowski form

$$ds^2 = -d\tau^2 + dr^2 + r^2 d\Omega_{(2)}^2 \tag{4.92}$$

---

[3] Here the overdot denotes a derivative with respect to an affine parameter along the null geodesic.

by the coordinate transformation [288]

$$\tau = t \cosh \chi, \tag{4.93}$$

$$r = t \sinh \chi. \tag{4.94}$$

4.6.2   Generalization to any static spherical geometry

The cosmological analogy based on timelike and null geodesics can be generalized to many static and spherically symmetric spacetimes, for which the line element can always be written in the form [1, 296]

$$ds^2 = -e^{-2\Phi(R)} \left(1 - \frac{2M(R)}{R}\right) dt^2 + \frac{dR^2}{1 - 2M(R)/R} + R^2 \, d\Omega_{(2)}^2 \tag{4.95}$$

where $T^\mu \equiv (\partial/\partial t)^\mu$ is the timelike Killing vector and $M(R)$ is the Misner–Sharp–Hernandez mass[4] [206, 283].

For radial timelike geodesics, the energy E of a particle of mass m and four–momentum $p^\mu = mu^\mu$ is conserved along the geodesic, according to $p_\mu T^\mu = -E$ hence, denoting with $\bar{E} \equiv E/m$ the energy per unit mass, we have

$$u^0 = \frac{dt}{d\tau} = \frac{\bar{E} \, e^{2\Phi}}{1 - 2M/R}. \tag{4.96}$$

The normalization of the four–velocity $u_\mu u^\mu = -1$ reads

$$g_{00} \left(\frac{dt}{d\tau}\right)^2 + g_{11} \left(\frac{dR}{d\tau}\right)^2 = -1 \tag{4.97}$$

and, substituting (4.96), we obtain $(dR/d\tau)^2$ and rewrite the result as the formal Friedmann equation

$$\left(\frac{1}{R} \frac{dR}{d\tau}\right)^2 = \frac{\bar{E}^2 \, e^{2\Phi}}{R^2} - \frac{1}{R^2} + \frac{2M(R)}{R^3}. \tag{4.98}$$

To proceed, note that many spherical spacetimes of physical interest exhibit the property $g_{tt} g_{RR} = -1$, equivalent to $\Phi \equiv 0$. This relation is associated with special physical and geometric properties discussed in Refs. [100, 228]. For such spacetimes, Eq. (4.98) becomes the Friedmann–like equation

$$\left(\frac{1}{R} \frac{dR}{d\tau}\right)^2 = \frac{(\bar{E}^2 - 1)}{R^2} + \frac{2M(R)}{R^3}. \tag{4.99}$$

---

[4]In spherical symmetry, the Hawking–Hayward quasilocal mass [201, 204] reduces to the Misner–Sharp–Hernandez mass [205].

The analogous universe has curvature index $K = 1 - \bar{E}^2$ and cosmic fluid with energy density

$$\rho = \frac{3}{4\pi} \frac{M(a)}{a^3}. \tag{4.100}$$

As usual, we impose the covariant conservation equation (1.187) to complete the analogy, obtaining the effective equation of state

$$P = -\rho - \frac{a}{3} \frac{d\rho}{da} = -\rho - \frac{a}{4\pi}\left(\frac{M'}{a^3} - \frac{3M}{a^4}\right) = -\frac{M'}{4\pi a^2} \tag{4.101}$$

or, in time–dependent form,

$$P = -\frac{M'(a)}{4\pi a^2} \equiv w(a)\,\rho. \tag{4.102}$$

For radial null geodesics with affine parameter λ the photon energy E is still conserved, $u_\mu T^\mu = -E$, and

$$\frac{dt}{d\lambda} = \frac{E\,e^{2\Phi}}{1 - 2M(R)/R}; \tag{4.103}$$

now the normalization $u_\mu u^\mu = 0$ gives the Friedmann–like equation

$$\left(\frac{1}{R}\frac{dR}{d\lambda}\right)^2 = \frac{E^2\,e^{2\Phi(R)}}{R^2}. \tag{4.104}$$

When $g_{tt}\,g_{RR} = -1$, equivalent to $\Phi \equiv 0$, the analogous universe satisfies the equation $H^2 = -K/a^2$ where $K = -E^2 < 0$, producing the Milne universe already encountered. If $\Phi \neq 0$, a less trivial analogy could present itself. In Appendix C.1 we report examples of cosmic analogies arising from radial timelike and null geodesics of static spherical geometries [152].

4.6.3   *An exact expanding relativistic sphere*

It is time to face the full general–relativistic problem of a fluid sphere expanding in empty space which, in the Newtonian analogy, corresponds to the massive particle sitting on the ball surface. It is then one of the fluid particles composing this sphere. The radius of the latter is obviously larger than its Schwarzschild radius. Particles composing a general fluid do not follow geodesics due to the pressure gradient $\nabla_\mu P$, except for a dust with zero pressure. Now one cannot ignore the fluid below the initial particle radius $R(0)$: the equation of motion for the boundary of the relativistic sphere must be written down and solved.

The simple Schwarzschild interior solution describing a static sphere with uniform density [404] is not relevant and must be generalized. What is needed is a solution describing a fluid sphere of uniform density that expands or contracts while keeping the energy density uniform. The interior geometry must match the Schwarzschild exterior. Several possibilities come to mind for the motion of the ball surface: could it be geodesic or is the normal to this surface a non–geodesic vector? Certain analytical solutions interpreted as time–dependent and asymptotically flat fluid spheres were found long ago by Vaidya [413]. They contain, as special cases, the Schwarzschild interior solution [404] and the Oppenheimer–Snyder [307] solution. However, the most useful analytical solutions for our problem are those found by Smoller and Temple [375]. They contain special cases describing our physical context and several results of previous literature, which is summarized below.

Ref. [274] found that the FLRW geometry is the unique solution of the Einstein equations describing the spherical and shear–free motion of an electrically neutral perfect fluid obeying an equation of state. As pointed out in Ref. [383], however, this theorem requires the extra hypothesis that the energy density be uniform, $\partial\rho/\partial r = 0$ (fortunately this is one of the assumptions in our problem). Then, the interior of our dynamical fluid sphere can only be FLRW. This theorem justifies *a posteriori* the use of a uniform fluid sphere in the Newtonian analogy. A non–uniform fluid sphere would correspond to an inhomogeneous universe in the analogy. Contracting spheres of fluid with pressure were studied by Thompson and Whitrow [401,402] and Bondi [53,54] in the different context of gravitational collapse forming black holes.

To proceed, the Newtonian analogy requires that the fluid composing the dynamical sphere be a dust, which is the only situation in which a *homogeneous* interior geometry can match the Schwarzschild exterior.

4.6.3.1 *Special case of the Smoller–Temple shock wave solution*

Smoller and Temple [375] studied analytical solutions of the Einstein equations describing a spherical shock wave that expands in a gas. The interior and exterior geometries match on a shell of zero thickness on which there is no matter, hence there is no jump in the matching conditions.

As noted, the requirement of uniformity for the interior geometry with $\rho = \rho(t)$ and $P = P(t)$, forces the line element inside the shock wave to be FLRW,

$$ds^2 = -dt^2 + a^2(t)\left(\frac{dr^2}{1 - Kr^2} + r^2 d\Omega_{(2)}^2\right), \qquad (4.105)$$

where all values of the curvature index K are possible. The exterior geometry outside the shock wave is that of a static and spherical Oppenheimer–Tolman solution (usually discussed in the description of stellar interiors), with line element

$$ d\bar{s}^2 = -B(\bar{r})\,dt^2 + \frac{d\bar{r}^2}{1 - 2M/\bar{r}} + \bar{r}^2 d\Omega_{(2)}^2 \qquad (4.106) $$

in spherical coordinates[5] $\left(t, \bar{r}, \vartheta, \varphi\right)$, where $M(\bar{r})$ is the mass inside the sphere of radius $\bar{r}$, given by

$$ \frac{dM}{d\bar{r}} = 4\pi\,\bar{r}^2 \rho(\bar{r}) \qquad (4.107) $$

while $\rho(\bar{r})$ denotes the energy density at radius $\bar{r}$ and

$$ \frac{B'(\bar{r})}{B(\bar{r})} = -\frac{2\,\bar{P}'(\bar{r})}{\bar{\rho} + \bar{P}} \,. \qquad (4.108) $$

$\bar{P}$ is the pressure outside the shell and a prime denotes differentiation with respect to $\bar{r}$. The mass $M(\bar{r})$ is nothing but the Misner–Sharp–Hernandez quasilocal mass [206, 283] at radius $\bar{r}$, which is defined in any spherically symmetric spacetime by

$$ 1 - \frac{2M_{\mathrm{MSH}}}{R} = \nabla^\mu R \nabla_\mu R \doteq g^{RR} \,, \qquad (4.109) $$

where R is the areal radius. The last equality is only valid if the areal radius is the radial coordinate, as in the line element (4.106)).

Smoller and Temple match interior and exterior solutions on the surface of the spherical shock wave by imposing the Darmois–Israel junction conditions [101, 226, 249]. There are no jumps in the first and second fundamental forms and no matter on the shell. For an outgoing shock wave, the junction conditions give [375]

$$ r\,\dot{a} = \sqrt{1 - Kr^2}\,\sqrt{1 - \Theta}, \qquad (4.110) $$

$$ \dot{r}\,a = \sqrt{1 - Kr^2}\,\frac{\sqrt{1 - \Theta}}{\gamma\Theta - 1}, \qquad (4.111) $$

where

$$ \Theta = \frac{1 - 2M/r}{1 - Kr^2}, \qquad (4.112) $$

$$ \gamma = \frac{\rho + \bar{P}}{\bar{\rho} + \bar{P}} \,. \qquad (4.113) $$

---

[5]The comoving coordinate r inside the sphere differs from the curvature (or Schwarzschild–like) radius $\bar{r}$ outside.

The simple coordinate transformation

$$\bar{r} = a(t)\, r \tag{4.114}$$

connects interior and exterior coordinates [375], which implies that the areas of two–spheres of symmetry change smoothly across the shock wave. In other words, the areal radius $R = a(t)\, r$ of the FLRW interior geometry matches the areal radius $\bar{r}$ of the exterior Schwarzschild one and the surface of the ball then comoves with its FLRW interior.

To be true to our physical problem, we impose that the sphere expands in an empty exterior, then $\bar{\rho} = \bar{P} = 0$ and the spherical and asymptotically flat exterior geometry is forced to be Schwarzschild by the Birkhoff theorem [435] (this special case of the outgoing shock wave is reported in [375]). This condition requires the pressure P to vanish inside the entire fluid sphere in order to match the vanishing pressure at the boundary, and then the interior fluid can only be dust.

The vanishing of the outside pressure $\bar{P} \to 0$ is equivalent to $\gamma \to \infty$; this limit implies [375] $r = r_0 = $ const. at the surface of the ball. The fluid energy density then scales as that of a dust,

$$\rho(t) = \frac{3M}{4\pi\, r_0^3\, a^3}\,. \tag{4.115}$$

The Friedmann equation at this boundary reduces to

$$\dot{a}^2 = \frac{2M}{r_0^3\, a} - K\,. \tag{4.116}$$

The result that only a uniform ball of dust can be matched to a Schwarzschild exterior is present in several works predating the one by Smoller and Temple, including those of McVittie [281], Bondi [53], Mansouri [272], Mashhoon and Partovi [274], and Glass [181].

The physics is summarized by stating that the surface of the fluid sphere expands in the surrounding Schwarzschild vacuum while comoving with its interior. Since $P = 0$, the fluid particles follow radial geodesics because of spherical symmetry and the unit normal to the surface of the comoving fluid sphere remains a timelike geodesic vector. Radial geodesic congruences are normal to the surface of the ball because the vorticity vanishes. The radial geodesics of the interior geometry join smoothly those of the Schwarzschild exterior because the surface of the fluid sphere comoves with its FLRW interior [375].

Smoller and Temple discuss all the possible spatial curvatures of the FLRW three–space inside the fluid sphere [375]. If $K > 0$, the well–known

Oppenheimer–Snyder solution describing the collapse of a ball of dust [307] is recovered for an ingoing shock wave (the signs of the right hand sides of Eqs. (4.110) and (4.111) are changed), while the ball boundary spans the cycloid

$$a(\eta) = \frac{1}{2}\left(1 + \cos\eta\right),$$ (4.117)

$$t(\eta) = \frac{1}{2\sqrt{K}}\left(\eta + \sin\eta\right).$$ (4.118)

(Here we impose the initial conditions $a(0) = 1, \dot{a}(0) = 0$ and we normalize the curvature index to $\kappa = 2M/r_0^3$ [375].) If, instead, $K < 0$, the solution $a(t)$ is given implicitly by [375]

$$\sqrt{a + a^2} - \frac{1}{2}\ln\left[1 + 2\left(a + \sqrt{a + a^2}\right)\right] = \sqrt{|K|}\,t$$ (4.119)

with initial condition $a(0) = 0$. Finally, if $\kappa = 0$ the scale factor and the comoving surface are given by the Einstein–de Sitter law

$$a(t) = \left(\frac{9\pi M}{2}\right)^{1/3} t^{2/3}.$$ (4.120)

The surface of the fluid sphere and the scale factor of its dust–dominated FLRW interior obey the Newtonian energy equation (4.69).

The analogy between the relativistic equation for the expansion of the fluid sphere described by $R(t)$ and the Newtonian equation (4.69) were the subject of a remark by Bondi [53], while Mashhoon and Partovi [274] mention explicitly this Newtonian analogy for FLRW universes.[6]

4.6.3.2   *Vaidya geometries*

Proceeding with our physical problem, one could conceive of reducing the fluid sphere to a spherical shell expanding in empty space at the speed of light [152]. This situation is described by the well–known Vaidya solutions [414]. The exterior geometry remains Schwarzschild by virtue of the Birkhoff theorem [435], whereas the matter composing the shell can only be a null dust expanding or contracting at the speed of light. The solution is then one of the best–known Vaidya geometries [414] and the null shell

---

[6]See Ref. [181, 383], which corrects certain errors in [272, 274].

moves along a null geodesic. Also in this context, there is no pressure because the latter does not appear in the energy–momentum tensor of a null dust

$$T_{\mu\nu} = \rho \, k_\mu k_\nu \,, \tag{4.121}$$

where k^μ is null and geodesic, $k_\mu k^\mu = 0$ and $k^\nu \nabla_\nu k^\mu = 0$.

Since the normal to the shell is tangent to a radial null geodesic, the cosmological analogy yields again the Milne universe. There is no point in considering radial timelike geodesics because a spherical shell expanding at the speed of light always hits a massive test particle moving radially and starting just above the ball. Considering radial null geodesics makes sense because the shell will never reach a radial outgoing photon emitted above it.

4.6.3.3 *Quasilocal energy, Ricci and Weyl tensors*

As is well known, the curvature tensor $\mathcal{R}_{\mu\nu\rho\sigma}$ splits into a Ricci part built out of the Ricci tensor $\mathcal{R}_{\mu\nu}$ and the Weyl tensor $C_{\mu\nu\rho\sigma}$ (the remaining part) [435] as

$$\mathcal{R}_{\alpha\beta\gamma\delta} = C_{\alpha\beta\gamma\delta} + g_{\alpha[\gamma}\mathcal{R}_{\delta]\beta} - g_{\beta[\gamma}\mathcal{R}_{\delta]\alpha} - \frac{\mathcal{R}}{3}\,g_{\alpha[\gamma}\,g_{\delta]\beta}\,. \tag{4.122}$$

The Weyl tensor is further decomposed into its electric and magnetic parts $E_{\alpha\beta}$ and $H_{\alpha\beta}$ with respect to a chosen timelike observer [126]. Of these, the electric part possesses a Newtonian analogue [126]; the magnetic part does not and contains the propagating degrees of freedom of the gravitational field.

Consider an observer with timelike four–velocity u^μ. We follow Ref. [40], which corrects a sign in the magnetic part of the Weyl tensor defined in [126]. Then, the electric and magnetic parts of the Weyl tensor are defined as

$$E_{\alpha\gamma}(u) = C_{\alpha\beta\gamma\delta}\,u^\beta u^\delta \,, \tag{4.123}$$

$$H_{\alpha\gamma}(u) = \frac{1}{2}\,\eta_{\alpha\beta\mu\nu}\,C^{\mu\nu}{}_{\gamma\sigma}\,u^\beta u^\sigma \,, \tag{4.124}$$

respectively, where $\eta_{\alpha\beta\gamma\delta} \equiv \sqrt{-g}\,\epsilon_{\alpha\beta\gamma\delta}$, $\epsilon_{\alpha\beta\gamma\delta}$ is the Levi–Civita symbol, and g is the determinant of the metric tensor $g_{\mu\nu}$, resulting in

$$\eta^{\alpha\beta\gamma\delta} = \eta^{[\alpha\beta\gamma\delta]} \,, \tag{4.125}$$

$$\eta^{0123} = \frac{1}{\sqrt{-g}} \,. \tag{4.126}$$

$E_{\mu\nu}$ and $H_{\mu\nu}$ are purely spatial tensors according to the notion of three–space of the observer u^{μ},

$$E_{\mu\nu}\, u^{\mu} = E_{\mu\nu}\, u^{\nu} = H_{\mu\nu}\, u^{\mu} = H_{\mu\nu}\, u^{\nu} = 0 \qquad (4.127)$$

and they are symmetric

$$E_{\mu\nu} = E_{(\mu\nu)}\,, \qquad H_{\mu\nu} = H_{(\mu\nu)} \qquad (4.128)$$

and trace–free,

$$E^{\mu}{}_{\mu} = H^{\mu}{}_{\mu} = 0\,. \qquad (4.129)$$

The Weyl tensor is then reconstructed from its electric and magnetic parts according to the relation [40, 126]

$$C_{\alpha\beta\gamma\delta} = \left(g_{\alpha\beta\rho\sigma}\, g_{\gamma dpq} - \eta_{\alpha\beta\rho\sigma}\, \eta_{\gamma dpq} \right) u^{\rho} u^{p}\, E^{\sigma q}$$

$$- \left(\eta_{\alpha\beta\rho\sigma}\, g_{\gamma\delta\mu\nu} + g_{\alpha\beta\rho\sigma}\, \eta_{\gamma\delta\mu\nu} \right) u^{\rho} u^{\mu}\, H^{\sigma\nu}\,, \qquad (4.130)$$

where

$$g_{\alpha\beta\rho\sigma} \equiv g_{\alpha\rho}\, g_{\beta\sigma} - g_{\alpha\sigma}\, g_{\beta\rho}\,. \qquad (4.131)$$

More explicitly, we have

$$C_{\alpha\beta\gamma\delta} = u_{\alpha} u_{\gamma}\, E_{\beta\delta} - u_{\alpha} u_{\delta}\, E_{\beta\gamma} - u_{\beta} u_{\gamma}\, E_{\alpha\delta} + u_{\beta} u_{\delta}\, E_{\alpha\gamma}$$

$$- \eta_{\alpha\beta\rho\sigma}\, \eta_{\gamma\delta\mu\nu}\, u^{\rho} u^{\mu}\, E^{\sigma\nu} - \eta_{\alpha\beta\rho\sigma}\, u_{\gamma} u^{\rho}\, H^{\sigma}_{\delta} + \eta_{\alpha\beta\rho\sigma}\, u^{\rho} u_{\delta}\, H^{\sigma}_{\gamma}$$

$$- u_{\alpha} u^{\mu}\, \eta_{\gamma\delta\mu\nu}\, H^{\nu}_{\beta} + u_{\beta} u^{\mu}\, \eta_{\gamma\delta\mu\nu}\, H^{\nu}_{\alpha}\,. \qquad (4.132)$$

By construction, the general–relativistic version of the uniform ball problem contains a Riemann tensor that reduces to its pure Ricci part in the interior of the fluid sphere and to its pure Weyl part outside. To see this, remember that all FLRW geometries are conformally flat, while the Weyl tensor $C_{\alpha\beta\gamma}{}^{\delta}$ is conformally invariant [435] and it always vanishes in FLRW. As a consequence, $\mathcal{R}_{\alpha\beta\gamma\delta}$ reduces to its Ricci part inside the dust sphere. In the exterior vacuum region, the Ricci tensor given by the Einstein equations

$$\mathcal{R}_{\mu\nu} = 8\pi \left(T_{\mu\nu} - \frac{T}{2}\, g_{\mu\nu} \right) \qquad (4.133)$$

vanishes identically and $\mathcal{R}_{\mu\nu\rho\sigma}$ reduces to its Weyl part. The Ricci and Weyl tensors switch roles at the surface of the dust sphere, where they are both discontinuous. As for their respective scalars $\mathcal{R}$ and $C_{\mu\nu\rho\sigma}C^{\mu\nu\rho\sigma}$, the Ricci scalar

$$\mathcal{R} = -\rho + 3P = -\rho \qquad (4.134)$$

jumps discontinuously to zero at the boundary, while the Weyl scalar which vanishes identically everywhere in the FLRW interior jumps to its Schwarzschild value

$$C_{\mu\nu\alpha\beta}\,C^{\mu\nu\alpha\beta} = \frac{48\,m^2}{r^6} \qquad (4.135)$$

there.

The Newtonian character of the quasilocal mass appearing in the Newtonian analogy problem is also relevant. The interior mass (4.107) matches the constant Schwarzschild mass m at the surface of the dust sphere. For any value of the curvature index K, the mass M contained in a sphere of areal radius $R \le R_0$ in the FLRW interior is

$$M^{(-)}(R) = \frac{4\pi\rho}{3}\,R^3 \,. \qquad (4.136)$$

This expression coincides with the Misner–Sharp–Hernandez/Hawking–Hayward quasilocal energy. In the limit $R \to a(t)\,r_0$ to the ball surface, which comoves with it FLRW interior, the mass $M^{(-)}$ becomes constant. This is possible only because the energy density scales as a dust, $\rho \sim 1/a^3$, which is necessary in order to match the constant Schwarzschild mass m. It is a general fact that the Misner–Sharp–Hernandez mass of a comoving sphere of radius R in any FLRW space evolves according to [134, 138]

$$\dot{M}_{\mathrm{MSH}} + 3H\,\frac{P}{\rho}\,M_{\mathrm{MSH}} = 0\,; \qquad (4.137)$$

then the constancy of M_{MSH} reflects the absence of pressure. In FLRW universes, the Misner–Sharp–Hernandez mass has the peculiarity that it depends on the energy density ρ but not on the pressure P; on the contrary, its time derivative $\dot{M}_{\mathrm{MSH}}$ depends on P but not on ρ.

Is this mandatory? In principle, a different equation of state $P = w\,\rho$ with $w = $ const. for the fluid, which causes the energy density to scale as $\rho(a) = \rho_0/a^{3(w+1)}$, could be compensated if the sphere expands in the non–comoving way $R_0(t) \simeq t^{w+1}$, keeping $M^{(-)}$ constant. The Misner–Sharp–Hernandez mass of a sphere of radius R_0 (not comoving in general) evolves according to [134, 138]

$$\dot{M}_{\mathrm{MSH}} + 4\pi\rho\,R_0^3\left[H\left(1 + \frac{P}{\rho}\right) - \frac{\dot{R}_0}{R_0}\right] = 0\,; \qquad (4.138)$$

therefore, M_{MSH} remains constant if $\dot{R}_0/R_0 = (w+1)H$ for $P = w\,\rho$ with $w = \mathrm{const}$. However, this choice makes the pressure discontinuous at the boundary, which no longer follows a timelike geodesic, and a test particle initially lying on it would detach immediately. Then, the radial timelike geodesics inside and outside, together with the areal radii R and $\bar{r}$, no longer match and the geometry becomes discontinuous, which is unphysical. A discontinuity in the pressure signals a material layer on the boundary, which has nothing to do with the original Newtonian problem.

The mass used in the Oppenheimer–Tolman exterior of the Smoller–Temple shock waves is the Misner–Sharp–Hernandez construct used in spherical fluid dynamics [206, 283]. The more general Hawking–Hayward quasilocal energy [201, 204] reduces to it in spherical symmetry [205] and, further to the Arnowitt–Deser–Misner mass if spacetime is asymptotically flat and the limit $R \to +\infty$ is taken. The literature contains several other, inequivalent, quasilocal energies (see [394] for a review). In our problem, the exterior Oppenheimer–Tolman geometry reduces to the Schwarzschild one and, as noted, the Misner–Sharp–Hernandez/Hawking–Hayward mass $M_{\mathrm{MSH}}^{(+)}$ reduces to the constant Schwarzschild mass m.

It is a general fact that the Hawking–Hayward quasilocal mass (with or without spherical symmetry) decomposes into a contribution coming from the matter energy–momentum tensor $T_{\mu\nu}$ plus a "pure gravity" term coming from the Weyl tensor. Furthermore, the latter is due solely to the electric part of the Weyl tensor [137].

In detail: the Hawking–Hayward mass enclosed by a two–surface $\mathcal{S}$ is defined as follows by these authors [201, 204]. Let the spacetime metric be $g_{\mu\nu}$, and consider a spacelike, embedded, compact, and orientable two–surface $\mathcal{S}$. Let $h_{\mu\nu}$ and $\mathcal{R}^{(h)}$ denote, respectively, the two–metric and the Ricci scalar induced by $g_{\mu\nu}$ on $\mathcal{S}$, while μ is the volume two–form on $\mathcal{S}$ and A is its area. The congruences of ingoing $(-)$ and outgoing $(+)$ null geodesics that emanate from $\mathcal{S}$ have expansions $\theta_{(\pm)}$ and shear tensors $\sigma_{\mu\nu}^{(\pm)}$. ω^μ denotes the projection onto $\mathcal{S}$ of the commutator of the null normal vectors to $\mathcal{S}$ (the anoholonomicity) [204]. Then, the Hawking–Hayward quasilocal mass contained in $\mathcal{S}$ is [201, 204]

$$M_{\mathrm{HH}} \equiv \frac{1}{8\pi}\sqrt{\frac{A}{16\pi}} \int_{\mathcal{S}} \mu \left(\mathcal{R}^{(h)} + \theta_{(+)}\,\theta_{(-)} - \frac{1}{2}\,\sigma_{\mu\nu}^{(+)}\,\sigma_{(-)}^{\mu\nu} - 2\,\omega_\mu\,\omega^\mu \right). \quad (4.139)$$

Given this definition, the decomposition of the Riemann tensor into Ricci and Weyl parts and the further splitting of the Weyl into its electric and magnetic parts with respect to an observer who has four–velocity u^μ parallel to the unit normal to $\mathcal{S}$ cause also the Hawking mass to split. The result is [137]

$$M_{\text{HH}} = \sqrt{\frac{A}{16\pi}} \int_{\mathcal{S}} \mu \left(h^{\mu\nu} T_{\mu\nu} - \frac{2T}{3} \right)$$

$$- \frac{1}{8\pi} \sqrt{\frac{A}{16\pi}} \int_{\mathcal{S}} \mu \, \eta_{\alpha\beta\rho\sigma} \, \eta_{\gamma\delta\mu\nu} \, h^{\alpha\gamma} \, h^{\beta\delta} u^\rho u^\mu \, E^{\sigma\nu} . \quad (4.140)$$

Here the "pure gravity" contribution to M_{HH} (the second integral in the right–hand side of Eq. (4.140)) comes solely from the electric part of the Weyl tensor $E_{\mu\nu}$. The magnetic part $H_{\mu\nu}$ of the Weyl does not contribute. In this sense, the Hawking mass is "Newtonian" [137].

In our particular problem of the relativistic generalization of the Newtonian ball, the two–surface $\mathcal{S}$ is a sphere of radius R or $\bar{r}$ and the decomposition (4.140) of the quasilocal mass assumes the simple form

$$M_{\text{MSH}}(R) = \frac{4\pi R^3}{3} \rho \, \theta_H \left(R_0 - R \right) + m \, \theta_H \left(\bar{r} - \bar{r}_0 \right) , \quad (4.141)$$

where R is the areal radius (which reads $R = a(t) r$ inside and $\bar{r}$ outside) and $\theta_H(x)$ denotes the Heaviside step function. In the dust interior, M_{MSH} coincides with the first term on the right–hand side. In the exterior, it coincides with the second term. Matching interior and exterior makes the two terms equal on the boundary between these regions and guarantees the continuity of M_{MSH}. The quasilocal energy remains always "Newtonian". We conclude that the existence of a Newtonian analogy for FLRW cosmology is possible because the FLRW spacetime has vanishing magnetic part of the Weyl tensor according to static observers.

4.6.3.4 *Symmetries of the Newtonian problem*

We can use the symmetries of the Einstein–Friedmann equations to explore corresponding symmetries of the Newtonian analogue. Restrict to a spatially flat FLRW universe: recalling that the operations

$$\rho \to \bar{\rho} = \bar{\rho}(\rho) , \quad (4.142)$$

$$H \to \bar{H} = \sqrt{\frac{\bar{\rho}}{\rho}} \, H , \quad (4.143)$$

$$P \to \bar{P} = -\bar{\rho} + \sqrt{\frac{\rho}{\bar{\rho}}} \left(P + \rho \right) \frac{d\bar{\rho}}{d\rho} \quad (4.144)$$

leave the $\kappa = 0$ Einstein–Friedmann equations invariant, in order for this transformation to be a symmetry of the Newtonian analogue it must be $P = \bar{P} = 0$. Then, using the new variable

$$z \equiv \frac{1}{\sqrt{\rho}}, \tag{4.145}$$

Eq. (4.144) becomes

$$\frac{d\bar{z}}{dz} = \frac{\bar{\rho}^{-3/2}}{\rho^{-3/2}} \frac{d\bar{\rho}}{d\rho} = 1 \tag{4.146}$$

and $\bar{z}(z) = z - z_0$, or

$$\frac{\bar{\rho}}{\rho} = \frac{1}{1 + \frac{\rho}{\rho_0} - 2\sqrt{\frac{\rho}{\rho_0}}} \tag{4.147}$$

with ρ_0 and z_0 integration constants. So, we learn that the specific map

$$\rho \to \bar{\rho} = \frac{\rho}{1 + \frac{\rho}{\rho_0} - 2\sqrt{\frac{\rho}{\rho_0}}}, \tag{4.148}$$

$$H \to \bar{H} = \frac{H}{1 + \frac{\rho}{\rho_0} - 2\sqrt{\frac{\rho}{\rho_0}}}, \tag{4.149}$$

preserves the condition $P = 0$ and the physics of the Newtonian fluid sphere encoded in Eq. (4.69) with $E = 0$.

For a dust–dominated and spatially flat FLRW universe we have $R(t) = R_0 \, t^{2/3}$, which yields

$$\bar{\rho}(t) = \frac{3M}{4\pi R_0^3} \frac{1}{t^2 - 2\sqrt{\frac{3M}{4\pi R_0^2 \rho_0}} \, t + \frac{3M}{4\pi R_0^3 \rho_0}}, \tag{4.150}$$

$$\overline{\left(\frac{\dot{R}}{R}\right)} = \left(\frac{\dot{R}}{R}\right) \frac{t^2}{t^2 - 2\sqrt{\frac{3M}{4\pi R_0^3 \rho_0}} \, t + \frac{3M}{4\pi R_0^2 \rho_0}}. \tag{4.151}$$

This Newtonian symmetry [152] does not appear in the pedagogical literature using uniform Newtonian spheres and gravity tunnels [8, 91, 92, 92, 104, 105, 109, 114, 161, 180, 196, 224, 225, 239, 242, 253, 270, 351, 362, 363, 373, 399, 415, 415, 447].

The last feature highlighted by the rather rich Newtonian analogy with FLRW cosmology consists of the fact that, for FLRW universes filled with a perfect fluid with constant equation of state $P = w\,\rho$, the Friedmann equation in conformal time reduces to a Riccati equation [132,233,343,345]. This property transfers to the Newtonian side of the analogy. Setting[7]

$$r = s^2 \,, \tag{4.152}$$

$$dt = r\, d\eta = s^2 d\eta \,, \tag{4.153}$$

reduces the Newtonian acceleration equation

$$\frac{d^2\vec{r}}{dt^2} = -\frac{GM}{r^3}\vec{r} \tag{4.154}$$

to

$$\frac{d^2\vec{s}}{d\eta^2} + \frac{|E|}{2}\vec{s} = 0 \,, \tag{4.155}$$

where

$$E = \frac{1}{2}\left(\frac{d\vec{r}}{dt}\right)^2 - \frac{GM}{r} < 0 \tag{4.156}$$

is the particle energy, changing the Coulomb force problem into that of two decoupled harmonic oscillators. This result was known to Euler [131] and keeps reappearing in the literature [25, 48, 52, 208, 258, 390].

---

[7]Equation (4.153) is analogous to the change from comoving to conformal time in FLRW universes.

Chapter 5

Cosmology on the beach

*Prediction by
analogy–creativity–is so
pervasive we normally don't
notice it.*

Jeff Hawkins

5.1 Equilibrium beach profiles

In oceanography, one of the most studied features of coastal morphology is
the equilibrium beach profile (Fig. 5.1). A beach profile is the depth of the
water, measured from the shore seaward and perpendicular to the shoreline,
a function of one variable. Beach profiles are important for coastal science
and very relevant for human activities [31, 45, 65, 102, 218, 219, 223, 246, 251,
252, 436]. Military operations motivated early research in beach profiles.

Beach profiles are subject to seasonal changes: typically, there are a
summer and a winter equilibrium profile. The summer equilibrium profile
is changed by fall and winter storms when materials are transported around
the sea bed and gradually the beach settles into a winter equilibrium profile
[445].

Researchers mostly study the simpler problem of equilibrium beach
profiles rather than following their dynamical changes [31, 45, 102, 218,
219, 223, 236, 246, 251, 252, 436]. The relatively abundant available data
[112,236,310,353] give the following picture: the landward side of the topog-
raphy increases for a while and there is an undulating relief, while the seaward side decreases [112,310,353]. Models of a beach profile match two different curves, one for the landward zone and one for the seaward zone, each
of which satisfies an appropriate ordinary differential equation [236,252].

Too often research in the earth sciences reduces to data collection and data–fitting rather than building theories and models for a general understanding, and this was the case for early research on beach profiles. Nowadays, however, there exist theories of beach profiles under different conditions, for example breaking or non–breaking waves at the shore. Larson *et al.* [252] and later Jenkins & Inman [236] have proposed a definite theory based on thermodynamics. Since, near the shore, wave motion causes turbulence and energy dissipation, the main ingredient of this theory [236, 252] consists of extremizing the rate of energy dissipation due to breaking and non–breaking waves.[1] Since the main ingredient is extremization, the theory can be formulated in terms of a variational principle for one–dimensional beach profiles. The resulting Euler–Lagrange equation is non-linear, making the problem non–trivial. It is rather surprising that this Euler–Lagrange equation lends itself to another cosmic analogy with a

---

[1] Both breaking and non–breaking waves cause energy dissipation but, following previous literature that focusses on the shorerise profile before wave breaking occurs, we will restrict to energy dissipation caused by friction at the bottom.

FIG. 5.1. Beach near Kannur, Kerala, India. The beach profile changes considerably with the monsoon due to the transport of sediments along the sea bed.

physically reasonable analogous universe [142].

Consider a horizontal x–axis perpendicular to the shore (assumed, for simplicity, to be straight) and pointing seaward. Let x be the cross–shore distance and $h(x)$ be the local water depth, measured vertically downward from a constant mean sea level.

The thermodynamic theory of Ref. [236, 252] identifies the equilibrium beach profile as the one that extremizes the energy dissipation rate. To this end, the functional [236, 252]

$$I\Big[x(h)\Big] = \int_{h_1}^{h_2} dh \Big[h(x)\Big]^{-\frac{3(n+1)}{4}} \sqrt{1 + \left(\frac{dx}{dh}\right)^2} \tag{5.1}$$

must be extremized, where $n > 0$ is an exponent appearing in the relation between shear stress amplitude τ_0 and water velocity $u_m(x)$ at the sea floor

$$\tau_0(x) = K_\tau \rho \Big[u_m(x)\Big]^n, \tag{5.2}$$

ρ is the density of seawater, and the constant K_τ is independent of u_m [236]. Relations of this kind are common in geophysical flows. The goal consists of determining the function $x(h)$ [236, 252].

For convenience, we recast the problem in terms of the actual depth profile $h(x)$, obtaining the different action functional of $h(x)$ [142]

$$J\Big[h(x)\Big] = \int_{x_1}^{x_2} dx \Big[h(x)\Big]^{-\frac{3(n+1)}{4}} \sqrt{1 + \left(\frac{dh}{dx}\right)^2}, \tag{5.3}$$

where the Lagrangian reads

$$L(h, h') = \Big[h(x)\Big]^{-\frac{3(n+1)}{4}} \sqrt{1 + (h')^2}, \tag{5.4}$$

and where a prime denotes differentiation with respect to x. Incidentally, the value $n = 7/3$ of the exponent, which is unphysical in this context [236], reproduces the variational principle and the Lagrangian

$$L = h \sqrt{1 + \left(\frac{dh}{dx}\right)^2} \tag{5.5}$$

of the catenary problem [42, 183] already studied in Chap. 2.

Since the Lagrangian (5.4) does not depend explicitly on x, the associated Hamiltonian

$$\mathcal{H} = p_h\, h' - L(h, h') \tag{5.6}$$

is conserved, where

$$p_h \equiv \frac{\partial L}{\partial h'} = \frac{h^{\frac{-3\,(n+1)}{4}} h'}{\sqrt{1 + (h')^2}} \tag{5.7}$$

is the momentum conjugated to h. The explicit expression of the Hamiltonian is

$$\mathcal{H} = -\frac{1}{h^{\frac{3\,(n+1)}{4}} \sqrt{1 + h'^2}} \tag{5.8}$$

and its conservation produces the first integral

$$h^{\frac{3\,(n+1)}{2}} \left(1 + h'^2\right) = C^2 \tag{5.9}$$

where $C \neq 0$ is an integration constant with dimensions

$$\left[C\right] = \left[L^{-\frac{3(n+1)}{4}}\right] \tag{5.10}$$

(if $C = 0$, the solution vanishes everywhere and is unphysical). By imposing the boundary condition $h(0) = 0$, *i.e.*, locating the origin where the water depth vanishes, one forces the derivative $h'(x)$ of the beach profile to diverge as $x \to 0$ to keep the left hand side of Eq. (5.9) finite.[2] Because of this cusp, the familiar existence and uniqueness theorems for the Cauchy problem at $x = 0$ [61] are not applicable.[3] As a consequence, the shallow water approximation used to formulate the theory of Ref. [236, 252] breaks down at the shore.

Proceeding in the usual way when we look for cosmic analogies, we manipulate Eq. (5.9) to bring it to the Friedmann–like form

$$\left(\frac{h'}{h}\right)^2 = \frac{C^2}{h^{\frac{3n+7}{2}}} - \frac{1}{h^2}, \tag{5.11}$$

which is the basis for the analogy. This form cannot be obtained if the variational problem for the beach profiles is not recast in terms of $h(x)$. The cosmological analogy [142] is not obvious, while the mechanical analogy with a point particle in one–dimensional motion is more straightforward.

---

[2] This boundary condition also has the effect of ruling out constant solutions, which are anyway unphysical.

[3] We will see in Chap. 6 that a similar situation occurs in the study of the one–dimensional longitudinal profile of a glacier ruled by the Vialov equation of glaciology—this profile has a cusp at the glacier terminus because the Vialov equation bears some similarity with Eq. (5.9) [143, 186, 216, 221, 323].

5.2 Point–particle mechanical analogy

The ordinary differential equation (5.9) can be rewritten as the energy conservation equation for a point particle of unit mass in one–dimensional motion along the h–axis

$$\frac{h'^2}{2} + V(h) = E, \tag{5.12}$$

where the potential energy is

$$V(h) = -\frac{C^2}{2\,h^{3(n+1)/2}} \tag{5.13}$$

and the total mechanical energy is fixed to the value $E = -1/2$. Here x plays the role of time. A qualitative understanding and a graphical representation of the possible solutions of Eq. (5.9) can be obtained from the graph of the potential $V(h)$ and its intersections with the horizontal line $E = -1/2$ in the Weierstrass approach [43, 44, 107, 183].

Only the region $h \geq 0$ is physically interesting and the potential energy $V(h)$ has a vertical asymptote at $h = 0$ and has the h–axis as a horizontal asymptote. Since $E \geq V$, the solutions $h(x)$ of Eq. (5.9) are always confined to the interval $0 \leq h \leq h_*$. The line $E = -1/2$ intersects the potential energy $V(h)$ at the turning point

$$\left(h_*, V_* \right) = \left(\left(\frac{C^2}{2|E|} \right)^{\frac{2}{3(n+1)}}, E \right) \tag{5.14}$$

which exists and is unique for all negative energies E, in particular for the relevant physical value $E = -1/2$. A solution $h(x)$ of Eq. (5.9) describes only the segment of the entire beach profile which lies seaward of the wave breaking point [112, 310, 353].

Since $V(h) \to -\infty$ as $h \to 0^+$, a particle coming from the right and approaching $h = 0$ has divergent kinetic energy in order for E to remain finite and constant at the value $-1/2$, reiterating the previous conclusion that $\lim_{h \to 0^+} h' = +\infty$. One can trace this divergent behaviour to the fact that Eq. (5.9) is obtained by Refs. [236, 252] under the approximation of a mild slope of the profile, expressed by the condition

$$\frac{\tan\beta}{k\,h} = \frac{h'}{k\,h} \ll 1, \tag{5.15}$$

where $\tan\beta = h'$ is the bottom slope and k is the wave vector of the water waves. The latter obeys the dispersion relation [236]

$$k \simeq \frac{\omega}{\sqrt{g\,h}}, \tag{5.16}$$

where ω is the angular frequency of the breaking wave and g is the acceleration of gravity. This implies that

$$\frac{h'}{k\,h} \simeq \frac{h'}{\sqrt{h}} \to \infty \tag{5.17}$$

as x and h approach zero and the mild slope approximation breaks down at the shore. Let us turn now to the analogy with FLRW cosmology.

5.3   Cosmic analogue of an equilibrium beach profile

The Friedmann–like equation (5.11) describes a $\kappa = +1$ FLRW universe under the formal identification of variables $\left(x, h(x)\right) \longrightarrow \left(t, a(t)\right)$. To make the analogy complete, it is sufficient that the covariant conservation equation (1.187) is satisfied, which yields the energy density

$$\rho(t) = \frac{\rho_0}{\left[a(t)\right]^{\frac{3n+7}{2}}} \tag{5.18}$$

of the analogous cosmic fluid, with $\rho_0 > 0$ an integration constant. Comparing with the FLRW energy density

$$\rho(a) = \frac{\rho_0}{a^{3(w+1)}} \,, \tag{5.19}$$

one obtains the equation of state parameter of the cosmic fluid filling the analogous FLRW universe

$$w = \frac{3n+1}{6} \,. \tag{5.20}$$

The requirement $n > 0$ in the model of Refs. [236, 252] then gives the restriction $w > 1/6$, which still allows for radiation ($w = 1/3$ and $n = 1/3$) and for a stiff fluid ($w = 1$ and $n = 5/3$). The latter corresponds to a free scalar field [82, 244, 260, 435]. The acceleration equation (1.180) predicts a decelerating analogous universe because $w > 1/6$.

5.4   Beach profile solutions via the analogous cosmology

The most interesting and useful part of the beach profile–FLRW universe analogy involves the solutions of the oceanography equation (5.11), which have been the subject of a controversy [142, 269].

5.4.1 *Case $n = 1/3$*

For $n = 1/3$, corresponding to a radiation fluid with $w = 1/3$, Eq. (5.18) expresses the typical blackbody scaling of the energy density $\rho(a) = \rho_0/a^4$. The scale factor of the analogous FLRW universe is the textbook solution [82, 244, 260, 435]

$$a(t) = \sqrt{C'} \sqrt{1 - \left(1 - \frac{t}{\sqrt{C'}}\right)^2}, \qquad (5.21)$$

where $C' > 0$ is an integration constant. The analogous FLRW universe has closed spatial sections, begins in a Big Bang singularity $a(0) = 0$, expands to the maximum size $\sqrt{C'}$, and ends its life in the Big Crunch $a\left(2\sqrt{C'}\right) = 0$. The corresponding equilibrium beach profile is

$$h(x) = h_0 \sqrt{1 - \left(1 - \frac{x}{h_0}\right)^2}, \qquad (5.22)$$

with h_0 a constant length. The graph of the equilibrium beach profile $h(x)$ in the interval $x \in \left(0, 2\,h_0\right)$ is a cycloid. This is the trajectory, in the vertical plane, of a point located on the rim of a wheel of radius h_0 rolling without slipping on the x–axis.

5.4.2 $n = -1$ *(linearly expanding universe)*

Although unphysical in the theory of Ref. [236, 252], other special values of the parameter n correspond to well known analytical solutions of FLRW cosmology and we comment on them anyway.

If $n = -1$ then $w = -1/3$, corresponding to the linear solution of the acceleration equation (1.180). The corresponding beach profile

$$h(x) = h_0\, x + h_1 \qquad (5.23)$$

has been considered in Ref. [269], while Ref. [103] reports it in the shallow water approximation. This linear solution is the only possible *exact* power–law solution of Eq. (5.9). (A wide variety of approximate solutions, on the contrary, can be power–law, see below.) In fact, assuming $h(x) = A\,x^\alpha$ with A and α constants, Eq. (5.9) readily gives $\left(n, \alpha\right) = \left(-1, 1\right)$ and $1 + A = \pm C$.

5.4.3 $n = -1/3$ *(dust–dominated cosmos)*

If $n = -1/3$, then $w = 0$, corresponding to cosmic dust and to another classic solution of FLRW cosmology given by the parametric representation [82, 244, 260, 435]

$$h(\eta) = \frac{C}{2}\left(1 - \cos\eta\right),\qquad(5.24)$$

$$x(\eta) = \frac{C}{2}\left(\eta - \sin\eta\right).\qquad(5.25)$$

Its expansion for $\eta \ll 1$ is

$$h(\eta) \simeq \frac{C}{4}\eta^2,\qquad(5.26)$$

$$x(\eta) \simeq \frac{C}{12}\eta^3,\qquad(5.27)$$

and

$$\frac{h}{x} \approx \frac{3}{\eta} \gg 1,\qquad(5.28)$$

which illustrates the meaning of the approximation $\eta \ll 1$: it corresponds to deep water. In this approximation the parameter η can be eliminated and then

$$h(x) \simeq \left(\frac{9\,C}{4}\right)^{1/3} x^{2/3}.\qquad(5.29)$$

This profile was claimed to fit well field data [103].

5.4.4   *General n and cosmic fluid*

For general values of the parameter n and of the cosmic fluid w, a solution of the Einstein–Friedmann equation (1.179)–(1.187) is found in parametric form up to a quadrature by performing a suitable variable change [250]. The Einstein–Friedmann equations written using conformal time η give

$$\eta = \pm \int \frac{da}{a\sqrt{\frac{8\pi}{3}\rho a^2 - K}}\qquad(5.30)$$

and, for $w = \text{const.}$,

$$\eta = \pm \int \frac{da}{a\sqrt{\frac{8\pi}{3}a^{-(3w+1)} - K}}.\qquad(5.31)$$

We now use the rescaled variable

$$z \equiv \left(\frac{8\pi C_1}{3} \right)^{\frac{-1}{3w+1}} a ; \qquad (5.32)$$

for curvature index $K = +1$ the relevant integral is

$$\int \frac{dz}{z\sqrt{z^m - 1}} = \frac{2}{m} \operatorname{arcsec} \left(z^{m/2} \right) , \qquad (5.33)$$

which allows one to integrate Eq. (5.31) and to invert the result, obtaining the solution in parametric form [87, 132, 250]

$$a(\eta) = a_0 \left[\cos \left(c\,\eta + d \right) \right]^{1/c} , \qquad (5.34)$$

$$t(\eta) = a_0 \int_0^\eta d\eta' \left[\cos \left(c\,\eta' + d \right) \right]^{1/c} , \qquad (5.35)$$

where

$$c = \frac{3w + 1}{2} \qquad (5.36)$$

and a_0 is a constant. The Big Bang boundary condition $a(0) = 0$ (with $t = 0$ corresponding to $\eta = 0$) determines the phase constant as $d = -\pi/2$ and then

$$a(\eta) = a_0 \left[\sin \left(\frac{3w + 1}{2} \eta \right) \right]^{\frac{2}{3w+1}} , \qquad (5.37)$$

$$t(\eta) = a_0 \int_0^\eta d\eta' \left[\sin \left(\frac{3w + 1}{2} \eta' \right) \right]^{\frac{2}{3w+1}} . \qquad (5.38)$$

On the beach profile side, the analogue of the conformal time is defined by $d\eta = dx/h(x)$. Infinitesimal increments of the dimensionless quantity η are infinitesimal increments of the distance from the shoreline measured in units of the local water depth. In finite terms, the analogue of Eq. (5.31) is

$$\eta = \pm \int \frac{dh}{h \sqrt{\frac{8\pi}{3} h^{\frac{-(3n+7)}{2}} - 1}} . \qquad (5.39)$$

Performing the integration and inverting the result gives [142]

$$h(\eta) = h_0 \left[\sin \left(\frac{3(n + 1)}{4} \eta \right) \right]^{\frac{4}{3(n+1)}} , \qquad (5.40)$$

$$x(\eta) = h_0 \int_0^\eta d\eta' \left[\sin \left(\frac{3(n + 1)}{4} \eta' \right) \right]^{\frac{4}{3(n+1)}} , \qquad (5.41)$$

with c as in Eq. (5.36).

The parameter value $n = 1/3$ considered in the previous subsection gives $c = 1$ and the integration of Eq. (5.41) becomes trivial. In this case the parameter η can be eliminated and the explicit solution (5.21) for $h(x)$ follows.

An alternative way to solve for the cosmic dynamics uses the acceleration equation and the fact that, in conformal time, the latter takes the form of a Riccati equation [132]. The acceleration equation (1.180) can be turned into

$$\frac{\ddot{a}}{a} + \frac{c\dot{a}^2}{a^2} + \frac{cK}{a^2} = 0. \tag{5.42}$$

Passing to conformal time η one obtains, for $K = +1$, the Riccati equation

$$\frac{1}{a}\frac{d^2a}{d\eta^2} + \frac{(c-1)}{a^2}\left(\frac{da}{d\eta}\right)^2 + c = 0, \tag{5.43}$$

which is solved [210, 222] using the auxiliary variable

$$u \equiv \frac{1}{a}\frac{da}{d\eta} \tag{5.44}$$

and setting

$$u \equiv \frac{1}{cv}\frac{dv}{d\eta}. \tag{5.45}$$

Then the Riccati equation (5.43) reduces to the harmonic oscillator equation

$$v'' + c^2 v = 0 \tag{5.46}$$

which has sine and cosine solutions. Returning to the original variable $a(\eta)$ gives back the solution (5.40) and (5.41) [132].

Another relevant question is when the solution can be expressed in terms of elementary functions, as it happens in the case $n = 1/3$ discussed above. This question is answered [87] by the Chebyshev theorem of integration [85, 273]. The Friedmann equation (1.179) in comoving time yields [87]

$$t = \int da \ \frac{a^{\frac{3w+1}{2}}}{\sqrt{\frac{8\pi\rho_0}{3} - a^{3w+1}}} \ ; \tag{5.47}$$

introducing [87]

$$b_0 \equiv \frac{8\pi\rho_0}{3}, \qquad u \equiv a^{\frac{3(w+1)}{2}}, \tag{5.48}$$

we have

$$t = \frac{2}{3(w+1)} \int \frac{du}{\sqrt{b_0 - u^\gamma}}$$

(5.49)

where

$$\gamma = \frac{2\,(3w+1)}{3\,(w+1)}$$

(5.50)

for $w \neq -1$. The Chebyshev theorem states that the integral is elementary if and only if $1/\gamma$ or $\frac{2 - \gamma}{2\gamma}$ is an integer [87]. If

$$\frac{1}{\gamma} = N = 0, \pm1, \pm2 \pm 3, \, ... \, ,$$

(5.51)

then $w = \dfrac{3 - 2N}{3(2N - 1)}$ and

$$n = \frac{7 - 6N}{3(2N - 1)}.$$

(5.52)

Then the physical requirement of [236, 252] that $n > 0$ gives $\frac{1}{2} < N < \frac{7}{6}$, which leaves only $N = 1$, corresponding to the radiation–dominated analogous universe with $n = w = 1/3$. The other possibility

$$\frac{2 - \gamma}{2\gamma} = N$$

(5.53)

yields

$$w = \frac{1 - N}{3N}$$

(5.54)

and $n = (2 - 3N)/(3N)$. The requirement $n > 0$ is then equivalent to $0 < N < 2/3$, which is not satisfied by any integer.

5.4.5 *Deep water approximation*

We can expand the general beach profile solution (5.40) and (5.41) in the deep water approximation $\eta \ll 1$. The result is

$$h(\eta) \simeq h_0 \left[\frac{3(n+1)}{4} \right]^{\frac{4}{3(n+1)}} \eta^{\frac{4}{3(n+1)}},$$

(5.55)

$$x(\eta) \simeq h_0 \left[\frac{3(n+1)}{4} \right]^{\frac{4}{3(n+1)}} \frac{3(n+1)}{7 + 3n} \eta^{\frac{7+3n}{3(n+1)}}.$$

(5.56)

The ratio

$$\frac{h}{x} \simeq \frac{(7+3n)}{3\,(n+1)} \frac{1}{\eta} \gg 1 \qquad (5.57)$$

does not depend on n and shows that $\eta \ll 1$ corresponds to deep water $h \gg x$. This deep water approximation is not the same used in wave mechanics, in which the depth of the water is instead much larger than the wavelength of the waves considered.

The elimination of the parameter η yields the approximate power–law beach profile

$$h(x) \simeq h_0^{\frac{3(n+1)}{7+3n}} \left(\frac{7+3n}{4}\right)^{\frac{4}{7+3n}} x^{\frac{4}{7+3n}}. \qquad (5.58)$$

Power–law functions were proposed since the early days of research on equilibrium beach profiles [57, 65, 103]. We recover the special case $n = -1/3$ already seen, for which the beach profile exponent is $2/3$, which was advocated in Ref. [103]. The other value $n = 1$ produces the exponent $2/5$ advocated in Ref. [57]. In general, the parameter n produces the exponent $4/(7+3n)$. One of these power–law solutions is recovered independently, with the same exponent, in Ref. [268].

5.4.6   *Beach profiles are roulettes*

The property of the Friedmann equation (1.179) that the graphs of all its solutions are roulettes [87] transfers to beach profiles. As already remarked, a roulette is the curve described by a point that lies on, or inside, a curve that rolls without slipping along a straight line.[4]

All the beach profiles proposed as analytical solutions of Eq. (5.11) in Ref. [236] are expressed by elliptic integrals of the first and second kind and have graphs that are elliptical cycloids, which are the curves described by a point on an ellipse when the latter rolls on the x–axis. If this ellipse degenerates into a circle, the elliptical cycloid reduces to an ordinary cycloid (a semi–circle, as in Eq. (5.22)). However, the derivation of these solutions in [236] is unclear and later authors were unable to reproduce them, which was the subject of criticism [268, 269].

The Friedmann equation (1.179) for a closed universe was studied in Ref. [87] to derive the equation of the solution in polar coordinates $\left(r, \vartheta\right)$. Referring the reader to [87] for details, we summarize its results. In general,

---

[4]In a more general definition, not of interest here, the curve rolls without slipping along another curve.

$r(\vartheta)$ is given up to a quadrature and not explicitly. There are, however, integrable cases when the cosmic fluid has energy density

$$\rho(a) = \frac{\alpha}{a^2} + \beta \, a^\delta \,, \tag{5.59}$$

where α, β, and δ are arbitrary constants [87] (but we require α and β to be non–negative to avoid negative densities). The situation of interest for beach profiles corresponds to

$$\alpha = 0 \,, \tag{5.60}$$

$$\beta = \rho_0 \,, \tag{5.61}$$

$$\delta = -\frac{(3n+7)}{2} \,. \tag{5.62}$$

The corresponding beach profile solution, constructed as a roulette in Ref. [87], is

$$\frac{1}{r^{\frac{3n+7}{3n+1}}} = \cos\left(\frac{3n+7}{3n+1}\vartheta\right) \,. \tag{5.63}$$

This expression contains simple solutions for $n = -1/3$ (corresponding to $w = 0$) [87]) and $n = -1/9$ (or $w = 1/9$), which are excluded by the physical requirement $n > 0$ of [236, 252].

The elliptical cycloid solutions of Ref. [236] are roulettes, but they are not reproduced by Eq. (5.63), adding to the criticism of Ref. [269]. To conclude, the oceanography side of the analogy gains quite a bit of insight from a century of knowledge on the cosmology side of the analogy, especially about analytical solutions of the equation ruling equilibrium beach profiles.

One can speculate about possible extensions of the relatively simple beach profile model used in the literature and adopted here, in parallel with other topics studied in cosmology [142]. For example, one could consider small perturbations of an equilibrium beach profile solution, which parallels the huge literature on perturbations of FLRW cosmologies.

5.5 Maximum or minimum energy dissipation?

There is a disagreement in the literature on whether the beach profiles extremizing the variational integral correspond to a maximum or a minimum of the energy dissipation rate. Since these are opposite points of view, one wonders how the same theory can accommodate mechanisms producing vastly different results for energy dissipation, a question investigated

in [150, 268]. Jenkins & Inman [236] advocate the second law of thermodynamics to support their view that the equilibrium beach profile corresponds
to a maximum of the dissipation rate. On the contrary, the original work
by Larson *et al.* [252] and the recent Ref. [269] disagree with this statement,
and even support the idea that the equilibrium beach profile corresponds
to a minimum of the energy dissipation rate instead. This question com-
pounds with the already mentioned criticism of the analytical solutions
of [236], which are disproved in [269]. The more comprehensive work [268]
supports the idea of a minimum, albeit with some *caveat* and subject to the
additional constraint that the bed slope angle does not exceed the angle of
repose of the sediments.

Although maximum dissipation rate and, according to intuition, maximum entropy appeal to physical intuition, the shorezone is a complex system. Larson *et al.* [252] argue that the flow at the sea bottom becomes
more efficient by minimizing energy dissipation, an argument encountered
elsewhere in the earth sciences in the presence of flows. Ref. [269] supports
indirectly the idea of a minimum, while Ref. [268] backs up this point of
view with field data.

The physical interpretation of a natural phenomenon should not be
an *a posteriori* rationalization but should rely on solid theoretical arguments. The issue of maximum or minimum entropy generation rates is a
big topic in the earth sciences, in particular in climate science, with no
general answer on whether entropy generation is maximum or minimum
(*e.g.*, [71, 108, 184, 188, 295]). Fortunately, in our situation it is easy to assess whether equilibrium beach profiles maximize or minimize the action
integral. Pragmatically, all that we need to do is computing the second
variation of this variational integral and determining its sign. Although
this procedure does not lead to definite results in many similar problems in
the earth sciences, it is straightforward for our particular problem [150].

We stress that, for the purpose of determining the beach profile, it
is immaterial whether the latter maximizes or minimizes the variational
integral because one only needs an extremum, and this is the reason why we
could proceed thus far without addressing the maximum versus minimum
problem. However, the solution to this problem is crucial to understand
the underlying physics.

The first variation of the action functional (5.3) determines the equation
already discussed for the equilibrium beach profiles. Consider variations
$\eta(x)$ that vanish at the endpoints, $\eta(-x_0) = \eta(x_0) = 0$ around the path
that extremizes the action (5.3). These variations are parametrized by a

parameter α. Accordingly, the varied paths are described by [438]

$$h(x,\alpha) = h(x,0) + \alpha\,\eta(x) \qquad (5.64)$$

where $h(x,0)$, corresponding to the parameter value $\alpha = 0$, is the path extremizing the action $J\big[h(x,\alpha)\big]$. Differentiation with respect to x gives

$$h'(x,\alpha) = h'(x,0) + \alpha\,\frac{d\eta}{dx} \qquad (5.65)$$

(where $f'(x) \equiv df/dx$), while differentiating with respect to the parameter α yields

$$\frac{\partial h}{\partial \alpha} = \eta\,, \qquad (5.66)$$

$$\frac{\partial h'}{\partial \alpha} = \frac{d\eta}{dx}\,, \qquad (5.67)$$

$$\frac{\partial^2 h'}{\partial \alpha^2} = \frac{\partial^2 h}{\partial \alpha^2} = 0\,. \qquad (5.68)$$

The use of these relations leads to the second variation of the action [438]

$$\frac{\partial^2 J}{\partial \alpha^2} = \int_{x_1}^{x_2} dx \left[\frac{\partial^2 L}{\partial h'^2} \left(\frac{d\eta}{dx}\right)^2 + 2\,\frac{\partial^2 L}{\partial h\,\partial h'}\,\eta\,\frac{d\eta}{dx} + \frac{\partial^2 L}{\partial h^2}\,\eta^2 \right]\,. \qquad (5.69)$$

Computing the derivatives appearing in this integral gives

$$\frac{\partial^2 L}{\partial h^2} = \frac{3\,(n+1)\,(3n+7)}{16}\,h^{-\frac{(3n+11)}{4}}\sqrt{1+h'^2}\,, \qquad (5.70)$$

$$\frac{\partial^2 L}{\partial h\,\partial h'} = -\frac{3\,(n+1)\,h'}{4\sqrt{1+h'^2}}\,h^{-\frac{(3n+7)}{4}}\,, \qquad (5.71)$$

$$\frac{\partial^2 L}{\partial h'^2} = \frac{h^{-\frac{3(n+1)}{4}}}{\left(1+h'^2\right)^{3/2}}\,, \qquad (5.72)$$

but they must be evaluated along the actual path that extremizes the action J. As already seen, these paths satisfy Eq. (5.9) and, substituting this information, the derivatives become

$$\frac{\partial^2 L}{\partial h^2} = \frac{3(n+1)(3n+7)C}{16} h^{-\frac{(3n+7)}{2}},\tag{5.73}$$

$$\frac{\partial^2 L}{\partial h\,\partial h'} = -\frac{3(n+1)}{4C}\frac{h'}{h},\tag{5.74}$$

$$\frac{\partial^2 L}{\partial h'^2} = \frac{h^{\frac{3(n+1)}{2}}}{C^3}.\tag{5.75}$$

Inserting these quantities into the integral $\partial^2 J/\partial\alpha^2$ gives

$$\frac{\partial^2 J}{\partial\alpha^2} = \int_{x_1}^{x_2} dx \left\{ \frac{h^{\frac{3(n+1)}{2}}}{C^3}\left(\frac{d\eta}{dx}\right)^2 - \frac{3(n+1)}{4C}\frac{h'}{h}\frac{d(\eta^2)}{dx} \right.$$

$$\left. +\frac{3(n+1)(3n+7)C}{16} h^{-\frac{(3n+7)}{2}}\eta^2 \right\}\tag{5.76}$$

$$\equiv I_1 + I_2 + I_3,\tag{5.77}$$

where the integral is now computed along the beach profiles $h(x)$ that extremize the action functional J. We now evaluate the signs of the three terms $I_{1,2,3}$. Useful information available is that $n > 0$, $C > 0$, and

$$h(x) \geq 0 \quad \forall x \in \left[x_1, x_2\right],\tag{5.78}$$

$$h'(x) > 0 \quad \forall x.\tag{5.79}$$

Begin with the first integral I_1 appearing in Eq. (5.77): its integrand is non–negative and $I_1 > 0$. Continue with the second integral I_2: integrating by parts and using the fact that the variations vanish at the endpoints, $\eta(x_1) = \eta(x_2) = 0$, gives

$$I_2 \equiv -\frac{3(n+1)}{4C}\int_{x_1}^{x_2} dx\,\frac{h'}{h}\frac{d(\eta^2)}{dx}$$

$$= -\frac{3(n+1)}{4C}\left\{\left[\eta^2\frac{h'}{h}\right]_{x_1}^{x_2} - \int_{x_1}^{x_2} dx\,\eta^2\left(\frac{h'}{h}\right)'\right\}$$

$$= \frac{3(n+1)}{4C}\int_{x_1}^{x_2} dx\,\eta^2\left(\frac{h'}{h}\right)'.\tag{5.80}$$

To compute the integrand, rewrite Eq. (5.9) in the Friedmann–like form

$$\left(\frac{h'}{h}\right)^2 = \frac{C^2}{h^{\frac{(3n+7)}{2}}} - \frac{1}{h^2},\tag{5.81}$$

differentiate it and divide term to term by $2h'/h$. The result is

$$\left(\frac{h'}{h}\right)' = -\frac{(3n+7)\,C^2}{4}\,h^{-\frac{(3n+7)}{2}} + \frac{1}{h^2}\,. \tag{5.82}$$

Now substitute Eq. (5.82) into I_2 and note that the first term on the right hand side of Eq. (5.82) cancels the integrand of I_3. We are left with

$$I_2 + I_3 = \frac{3\,(n+1)}{4C}\int_{x_1}^{x_2} dx \left(\frac{\eta}{h}\right)^2 > 0\,. \tag{5.83}$$

Adding $I_{1,2,3}$ gives

$$\frac{\partial^2 J}{\partial \alpha^2} = I_1 + I_2 + I_3$$

$$= \frac{1}{C^3}\int_{x_1}^{x_2} dx\, h^{\frac{3(n+1)}{2}}\left(\frac{d\eta}{dx}\right)^2 + \frac{3\,(n+1)}{4C}\int_{x_1}^{x_2} dx \left(\frac{\eta}{h}\right)^2\,. \tag{5.84}$$

We know that h, C, and n are positive and that the variations η and their derivatives cannot vanish simultaneously everywhere, hence the two contributions to the right hand side of Eq. (5.84) are positive and

$$\frac{\partial^2 J}{\partial \alpha^2} > 0\,. \tag{5.85}$$

The equilibrium beach profile $h(x,0)$ extremizing J corresponds to a *minimum* of the action functional J. This conclusion disproves the interpretation of Jenkins & Inman [236] and confirms that of Larson *et al.* [252]. It agrees with Ref. [268], which provides substantial physical evidence that, under the assumptions and approximations made, the equilibrium beach profile minimizes the action, although it does not compute $\partial^2 J/\partial \alpha^2$.

The interpretation of this result is less clear. Jenkins & Inman [236] advocate for a maximum on the basis of the second law of thermodynamics, *i.e.*, maximization of entropy in an isolated system (the shorezone system plus its surroundings [236]). Larson *et al.* [252] state that "the profile shape seaward of the break point at equilibrium is such that the waves dissipate a minimum of energy when travelling across the profile" and that "the waves lose minimum energy as they propagate across the profile, experiencing the least reduction in wave height for all possible profile shapes" [252]. This argument is plausible, although it is not tested directly. Another possible interpretation [150] is that the variational principle $\delta J = 0$ is designed to provide only the final state of equilibrium and not the approach to equilibrium. Once the final equilibrium beach profile is attained, entropy has already been maximized and the dissipation rate is minimum because the system cannot attain a higher entropy state. This idea is corroborated by the observation of Ref. [252] that wave dissipation is near–equivalent to the capacity of sediment transport through bottom friction. Then, from a physical perspective, it is fair to assume that sediment transport is at a minimum as well at equilibrium, otherwise there would be a dynamic equilibrium profile. Hence, the bottom friction and the wave energy dissipation must be minimum at equilibrium.

Chapter 6

Cosmology on a glacier

All perception of truth is the detection of an analogy.

Henry David Thoreau

6.1 Introduction

Alpine glaciers and polar ice caps are spectacular rivers of ice in slow flow (Fig. 6.1). Although they have been known for most of the human history, they have been avoided by people until the end of the 1700s and their exploration is relatively recent. Their mathematical modelling is even more recent, originating in the 1950s, although some attempts at their quantitative description had been made earlier [378–381].

On glaciers, snow reduces to ice under its own weight through a densification process. The growth of the average size of ice grains as the snow gets compacted presents a formal analogy with the Friedmann equation, although the corresponding first order differential equation does not describe an energy conservation as in the usual mechanical analogies that we have already seen between Friedmann equation and point particles in one–dimensional motion.

Glacial morphology can be quite obvious to recognize even thousands of years after the glacier has melted away. In this chapter we discuss glacial valleys, the transversal profiles of which are largely the imprint of the glaciers that carved them. We then discuss the longitudinal profile of a glacier or ice cap, that is, its "shape" along the direction of its flow. It is rather surprising that both glacial valley profiles (the transversal glacier profiles) and longitudinal glacier profiles can be related to cosmol-

FIG. 6.1. Glaciers flowing from Mt. Athabasca (3,491 m) in the Canadian Rocky Mountains range.

ogy through analogies, as pointed out in [86, 87] and discussed in detail in Refs. [149, 154].

6.2   Growth of grains in glacier firn

Snow falling on a glacier becomes more and more compact under the weight of the layers accumulating above and, eventually, it is compacted and becomes glacier ice. This process can be relatively fast, as in alpine glaciers, or it may take a very long time, up to a thousand years in the cold and dry polar regions [97, 323]. In alpine or coastal regions, the presence of liquid water accelerates this process. There are several complicated mechanisms in this compactification process, referred to in glaciology as the densification of snow [97, 323]. The growth of the grain size in the glacial firn[1] is an indicator of the progress of this densification. The relevant physical variable is the average size of the ice grains in the firn, defined by an effective grain diameter D [97]. Under the weight of the snow and firn above, the

---

[1]Glacial firn is a medium in between snow and ice: it is defined as partially consolidated snow that has experienced at least one summer melting season, but is not yet glacier ice. The density of firn ranges between 400 kg/m^3 and 830 kg/m^3 [97, 323].

grain boundaries tend to expand and D grows in time until the density of ice is reached. However, the situation is complex because of other factors playing a role that opposes grain growth. Realistically, ice contains air bubbles and impurities in the form of foreign particles that tend to slow down the grain growth by replacing the bonds between the water molecules in the solid state with weaker bonds with other chemical species. To simplify the problem, glaciologists assume dry snow conditions and uniform temperature throughout the firn. Let γ_{GB} be the energy per surface area of the grain boundary, which plays a role analogous to the surface tension at the surface of a liquid, and introduce an effective mobility of the grain boundary M^*. Let $R_s^{(b)}$ and $R_s^{(p)}$ be the effective radii of bubbles and particle impurities in the ice, respectively [97, 323]. All these quantities are considered to be constant [97, 323]. Then, the average size $D(t)$ of the ice grains increases with the time t according to [97]

$$\frac{dD}{dt} = c\,\gamma_{\mathrm{GB}}M^* \left(\frac{2}{D} - \frac{1}{R_s^{(b)}} - \frac{1}{R_s^{(p)}} \right), \tag{6.1}$$

where c is a phenomenological coefficient. The negative signs with which the effective radii $R_s^{(b)}$ and $R_s^{(p)}$ appear in Eq. (6.1) make it clear that bubbles and impurities delay the expansion of the ice grain boundaries and reduce their growth.

The form of Eq. (6.1) is the same as the form of Eq. (2.14) considered in Chap. 2 and in Ref. [144], or

$$\dot{D} = \frac{\alpha}{D} - \beta \equiv f(D), \tag{6.2}$$

with

$$\alpha = 2\,c\,\gamma_{\mathrm{GB}}M^* > 0, \tag{6.3}$$

$$\beta = c\,\gamma_{\mathrm{GB}}M^* \frac{\left(R_s^{(b)} + R_s^{(p)} \right)}{R_s^{(b)} R_s^{(P)}} > 0. \tag{6.4}$$

The corresponding Lagrangian (2.27), discussed in Chap. 2, takes the form [144]

$$L = \dot{D}^2 + \left(\frac{\alpha}{D} - \beta \right)^2; \tag{6.5}$$

as L does not depend on time, the associated Hamiltonian

$$\mathcal{H} = \dot{D}^2 - \left(\frac{\alpha}{D} - \beta \right)^2 \tag{6.6}$$

is conserved. Setting $\mathcal{H}$ to a constant does not reproduce Eq. (6.2) unless this constant is chosen to vanish, $\mathcal{H} = 0$. As discussed in Chap. 2, the constant

$$D(t) = D_* = \frac{\alpha}{\beta} = \frac{2\,R_s^{(b)}\,R_s^{(p)}}{R_s^{(b)} + R_s^{(p)}} \tag{6.7}$$

is an exact solution of the first order differential equation. It corresponds to $f(D_*) = 0$ or, the grain size is the double of the "baricentre" of the radii $R_s^{(b)}$ and $R_s^{(b)}$ of the impurities considered, bubbles and foreign particles. One can rewrite the function $f(D)$ as

$$f(D) = \beta \left(\frac{D_*}{D} - 1 \right). \tag{6.8}$$

As already discussed in Chap. 2, D_* is a single root of $f(D)$ with $f'(D_*) < 0$, therefore the constant solution (6.7) is stable and is a phase space attractor. Indeed, it is straightforward to check this statement by inspecting Eq. (6.2):

- If $D(t) > D_*$, then $\dot{D} < 0$ and $D(t)$ always decreases and approaches the value D_* as time progresses.
- If, instead, $D(t) < D_*$, then $\dot{D} > 0$ and $D(t)$ must increase, and it approaches the value D_* again.

The second possibility is the realistic situation and corresponds to $f(D) < 0$ as, during the densification of snow, the ice grains grow from a small size until bubbles disappear and the firn reaches the density of glacier ice.

It is easy to integrate Eq. (6.2), which yields the solution

$$t(D) = t_0 + \frac{1}{\beta} \left(D_* - D - D_* \ln |D - D_*| \right), \tag{6.9}$$

with t_0 an integration constant. The limit $D \to D_*$ in Eq. (6.9) corresponds to the limit $t \to +\infty$, illustrating again the fact that $D = D_*$ is the late-time attractor in phase space.

The grain growth process lends itself to a formal analogy with FLRW cosmology. Dividing Eq. (6.2) by D and squaring gives

$$\left(\frac{\dot{D}}{D} \right)^2 = \frac{\alpha^2}{D^4} + \frac{\beta^2}{D^2} - \frac{2\alpha\,\beta}{D^3}; \tag{6.10}$$

the first term on the right hand side corresponds to a radiation fluid, the second term to a curvature contribution $-K/a^2$ in the analogous Friedmann equation (1.179) with $K < 0$, while the third term corresponds to a dust with negative energy density. This feature is unphysical, but a formal analogy remains and makes it possible to balance the first two terms to achieve a static analogous universe corresponding to $D = $ const., which is an attractor of the dynamics. Of course, fluids with negative energy densities are not considered in FLRW cosmology, but this example shows what their effects would be. A static solution of the Einstein–Friedmann equations (1.179)–(1.187), the unphysical Einstein static universe is well known in cosmology [120, 435]. It corresponds to positively curved $(K > 0)$ spatial sections instead, to a positive cosmological constant Λ, and it contains a dust with positive energy density, and it is unstable [120, 435]. Curiously, swapping the signs of the curvature and of the dust energy density creates a static universe which, contrary to the Einstein static universe, is stable and is a phase space attractor.

6.3   The universe in a glacial valley

Glacial valleys are some of the most noticeable features of glacial morphology (Fig. 6.2). An old adage of geography says that valleys carved by rivers are V–shaped and valleys carved by glaciers are U–shaped. U–shaped valleys have been attributed to the carving and grinding of glaciers since the early days of glaciology [74, 279].

For many years, glaciologists contented themselves with fitting field data of glacial valley profiles with parabolas $y(x) = ax^2 + bx + c$ (a practice initiated in Ref. [393] and heavily criticized [327]). Other curves such as the power–law $y(x) = ax^b$, possibly with a different exponent b for the half–profile on each side of the centre $x = 0$, have been used. A parabola being the second order Taylor expansion of any sufficiently regular function with a minimum, this data–fitting is meaningless from the theoretical point of view. Furthermore, mere data–fitting does not contribute to understanding the process generating glacial valley profiles.

With the advent of theoretical modelling, the power of modern computers allows for the detailed numerical simulation of the dynamical process shaping a valley through the erosion of the valley walls and bed by a glacier over time [198, 370, 448]. A simple analytical description of the final product of this glacial process is also possible, and is obtained through a variational principle. The key idea, originally due to Hirano & Aniya [212] consists of extremizing the friction of the ice against the valley walls and bed, while enforcing the conservation of the ice mass during the process through a Lagrangian constraint in the corresponding variational principle.

Following Ref. [212], we describe the cross–sectional profile of a glacial valley with a function $y(x)$ of one variable, where x is a coordinate transverse to the glacier flow. According to Hirano & Aniya [212], friction should be minimum at the end of the erosion process with the isoperimetric constraint that the contact length of the cross–profile of the ice remain constant. Denoting differentiation with respect to x with a prime, the contact length of the ice between two endpoints x_1 and x_2 along the transverse profile is

$$\ell_{12}\Big[y(x)\Big] = \int_{x_1}^{x_2} d\ell = \int_{x_1}^{x_2} \sqrt{dx^2 + dy^2} = \int_{x_1}^{x_2} \sqrt{1 + (y')^2}\, dx = \text{const.}$$
(6.11)

The Coulomb model of friction is adopted, according to which the friction force is

$$f = \mu N,$$
(6.12)

where $N = \rho g h A_1$ is the normal force, μ is a friction coefficient, ρ is the ice density, g is the acceleration of gravity, h is the ice thickness, and A_1 is the area of contact between the ice and the bed. If water is present, its

FIG. 6.2.   A U–shaped valley carved by glaciers in British Columbia, Canada, with Sherbrooke Lake in the background.

pressure between the ice and the glacier bed is assumed to be constant and does not contribute to the variational principle [212].

Consider a unit width of ice in the longitudinal direction of the glacier: the friction force due to the infinitesimal contact length $d\ell$ is $df = \mu P d\ell$. Further, $P = \eta (y_s - y)$, where y_s denotes the ice surface and η is constant. Integrating along the entire transverse profile gives the total friction

$$f\Big[y(x)\Big] = \mu \, \eta \int_{x_1}^{x_2} \Big(y_s - y\Big) \sqrt{1 + (y')^2} \, dx \, . \qquad (6.13)$$

Extremizing this integral and implementing the condition (6.11) as a Lagrangian constraint gives the variational principle

$$\delta J = \mu \, \eta \, \delta \int_{x_1}^{x_2} dx \Big(y_s - y + \lambda\Big) \sqrt{1 + (y')^2} \equiv \delta \int_{x_1}^{x_2} dx \, L = 0 \, , \qquad (6.14)$$

where λ is a Lagrange multiplier and, apart from irrelevant multiplicative constants,

$$L = \Big(y_s - y + \lambda\Big) \sqrt{1 + (y')^2} \qquad (6.15)$$

is the Lagrangian. Since L does not depend explicitly on x, the corresponding Hamiltonian is conserved, giving [212]

$$\frac{y_s - y + \lambda}{\sqrt{1 + (y')^2}} = c_1 \, , \qquad (6.16)$$

where c_1 is a constant. We have already encountered this equation in the catenary problem of mechanics in Chap. 2, hence it comes to no surprise that the solutions found by Hirano & Aniya are catenaries [212]. Their method and results have been criticized by Harbor [197], originating a debate [197, 213, 214, 286]). It is however, established that, although the original idea was clever, the implementation was incorrect. The assumptions of Hirano & Aniya's model [212] seem to be inconsistent with common assumptions in glaciology at the time [197]. More important, friction should be maximized, not minimized according to Harbor [197]. It is true, however, that this criticism does not change the first order variational principle, in which friction is extremized. While Hirano & Aniya came to agree on this point [213], they stood by the validity of their result. It was only fifteen years later that the subject was reanalyzed by Morgan [286], who pointed out that the isoperimetric constraint (6.11) has no physical basis. One should instead require that the area of the cross–section of the glacial valley remains fixed. The physical meaning of this new constraint is that the volume of the (incompressible) ice is kept constant, because we are considering a unit width of ice in the direction of longitudinal flow [197]. In their reply to Morgan [214], Hirano & Aniya seem to have originally thought of this constraint, but it was discarded in their original paper [213].

By replacing Hirano & Aniya's constraint with the new one by Morgan, implemented again as a Lagrangian constraint in the action, the corrected variational principle leads to the Morgan equation [286]

$$\left(\frac{y'}{y}\right)^2 = \frac{1}{(\lambda y - C)^2} - \frac{1}{y^2}, \tag{6.17}$$

where now $y(x)$ is the ice thickness at transverse coordinate x, the valley profile (*i.e.*, the elevation of the glacier bed) is (see Fig. 6.3)

$$z(x) = y_s - y(x), \tag{6.18}$$

λ is a Lagrange multiplier, and C is a constant. In order to have a smooth symmetric solution $y(x)$ on the interval $\left[-x_0, x_0\right]$ with $y'(0) = 0$, it is necessary that $\lambda > 1$ and $C > 0$ [286]. The Morgan equation (6.17) is adopted to model the transverse profiles of glacial valleys.

Morgan [286] provided also the analytical solution solution of Eq. (6.17)

$$\left(\lambda^2 - 1\right)|x| = C\lambda\sqrt{1 - w^2} + C\arccos w, \tag{6.19}$$

where

$$w = \left(\frac{\lambda^2 - 1}{C}\right)y - \lambda, \tag{6.20}$$

$$-\frac{1}{\lambda} \leq w \leq 1, \tag{6.21}$$

$$\frac{C}{\lambda} \leq y \leq \frac{C}{\lambda - 1}. \tag{6.22}$$

In a different context, another formal solution of Eq. (6.17) for $\lambda^2 < 1$ is provided by Ref. [86],

$$\pm\frac{(1 - \lambda^2)^{3/2}}{|D|}x = \lambda\sqrt{w^2 - 1} + \ln\left|w + \sqrt{w^2 - 1}\right| + D, \tag{6.23}$$

where D is an integration constant and $|w| > 1$. For $C = 0$ and $|\lambda| < 1$ the solutions become linear.

The requirement of fixed cross–sectional area of the glacial valley appears also in the numerical simulations of the erosion process creating the U–shaped valley [370]. It is clear that a constraint is necessary because, if the friction integral is extremized without constraints, its first order variation

$$\delta \int_{x_1}^{x_2} \left(y_s - y \right) \sqrt{1 + (y')^2} \equiv \delta \int_{x_1}^{x_2} L_0 = 0 \qquad (6.24)$$

produces an Euler–Lagrange equation which has only unphysical solutions [141, 154]. In fact, the corresponding Lagrangian would be

$$L_0 \left(y(x), y'(x) \right) = (y_s - y) \sqrt{1 + (y')^2}, \qquad (6.25)$$

and the corresponding Hamiltonian

$$\mathcal{H}_0 = p_y \, y' - L_0 \qquad (6.26)$$

is conserved because (6.25) is independent of x. Here

$$p_y = \frac{\partial L_0}{\partial (y')} = \frac{(y_s - y) \, y'}{\sqrt{1 + (y')^2}} \qquad (6.27)$$

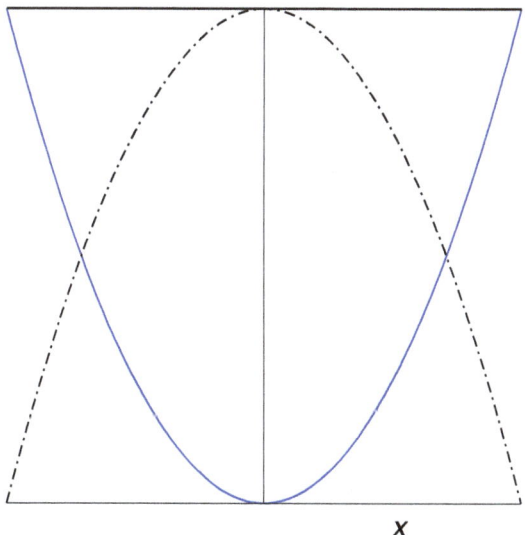

Legend

—·— **ice thickness** ——— **bed elevation**

FIG. 6.3. The transverse coordinate x spans the glacial valley, the ice thickness $y(x)$ is maximum at $x = 0$ where $y(0) = y_s$, and the bed elevation is $z(x) = y_s - y(x)$: this is the glacial valley profile.

is the momentum conjugated to y. The first integral reads

$$\frac{(y_s - y)(y')^2}{\sqrt{1 + (y')^2}} - (y_s - y)\sqrt{1 + (y')^2} = C, \qquad (6.28)$$

where C is a constant. By introducing the auxiliary variable $\xi \equiv y_s - y > 0$, Eq. (6.28) becomes

$$C\sqrt{1 + (\xi')^2} + \xi = 0, \qquad (6.29)$$

which requires that $C < 0$, hence we set $C \equiv -C_2$ where $C_2 > 0$ has the dimensions of a length. Since ξ is positive, it is

$$\xi' = \pm\sqrt{\frac{\xi^2}{C_2^2} - 1} \qquad (6.30)$$

which requires $\xi \geq C_2 > 0$. Then, all solutions $\xi(x)$ of Eq. (6.30) are unbounded above and are

$$y_s - y(x) = C_2^2 \, e^{\mp\frac{(x-x_0)}{C_2}} + \frac{e^{\pm\frac{(x-x_0)}{C_2}}}{4} \qquad (6.31)$$

where x_0 is another integration constant, and imply that $|x| \geq C_2$: this does not represent a valley geometry.

6.3.1   *Mechanical analogy for glacial valleys*

Assuming $C > 0$ [286], the Morgan equation (6.17) is rewritten as the energy conservation equation

$$\frac{(y')^2}{2} + V(y) = E \qquad (6.32)$$

for a fictitious point particle of unit mass and kinetic energy $(y')^2/2$ moving in one–dimensional motion along the y–axis in time x. The potential energy is

$$V(y) = \frac{-y^2}{2(\lambda y - C)^2} \qquad (6.33)$$

and the mechanical energy of this particle has the constant value $E = -1/2$. Newton's second law of motion for this particle is $y'' = -dV/dy$. The possible motions $y(x)$ of this particle can describe glacial valley profiles. A qualitative understanding of these motions is gained by examining the intersections between the potential energy $V(y)$ and the horizontal line $E = -1/2$, in the Weierstrass approach [43, 44, 107, 183].

The potential energy has the properties that $V(y) < 0 \ \forall y \neq 0$, $V(0) = 0$, and $V(y) \rightarrow -1/\left(2\lambda^2\right)$ as $y \rightarrow \pm\infty$. Furthermore, $V(y)$ has a vertical asymptote $y = C/\lambda > 0$ and

$$V'(y) = \frac{Cy}{(\lambda y - C)^3} \tag{6.34}$$

tells us that $V(y)$ increases for $y < 0$ and for $y > C/\lambda$, and decreases in the range $0 < y < C/\lambda$, with a maximum at $y = 0$.

Since a glacier has finite thickness, only bounded motions $y(x) > 0$ are meaningful, and we confine ourselves to the situation $C > 0, \lambda > 0$ which places the vertical asymptote $y = C/\lambda$ of $V(y)$ in the region of positive y. Equation (6.17) and the potential energy $V(y)$ are invariant under the exchange $(C, \lambda) \rightarrow (-C, -\lambda)$ so, at least formally, the situation $C < 0, \lambda < 0$ is the same as the one that we describe below.

6.3.1.1 *Parameter range $\lambda > 1$*

Consider $\lambda > 1$: if

$$E = -\frac{1}{2} < -\frac{1}{2\lambda^2}, \tag{6.35}$$

the horizontal line $E = -1/2$ describing the conserved energy lies below the horizontal asymptote of $V(y)$. Then, two regions correspond to bounded motions $y(x) > 0$: the first one is

$$0 < y_1 \leq y(x) < \frac{C}{\lambda}, \tag{6.36}$$

and the second region is

$$\frac{C}{\lambda} < y(x) \leq y_2, \tag{6.37}$$

where $y_{1,2}$ denote the turning points. The condition $\lambda > 1$ necessary for the existence of Morgan's bounded solutions [286] can then be interpreted graphically. The fictitious particle approaches the vertical asymptote $y = C/\lambda$ of $V(y)$ and is confined to either the region (6.36) or the region (6.37).

At the turning points $y_{1,2}$ (which correspond to zero slope of the valley profile $y(x)$ and to its lowest point), the kinetic energy $(y')^2/2$ vanishes which, using (6.17), yields the quadratic

$$\left(\lambda^2 - 1\right)y^2 - 2\lambda C y + C^2 = 0 \tag{6.38}$$

with roots

$$y_{1,2} = \frac{C}{\lambda \pm 1}. \tag{6.39}$$

The range

$$\frac{C}{\lambda} < y(x) \le \frac{C}{\lambda - 1} \qquad (6.40)$$

reproduces the condition (6.22) of Ref. [286], while the other interval

$$\frac{C}{\lambda + 1} \le y(x) < \frac{C}{\lambda} \qquad (6.41)$$

was not considered there.

6.3.1.2   *Parameter $\lambda = 1$*

For the special value $\lambda = 1$ of the Lagrange multiplier, corresponding to $E = -\frac{1}{2} = -\frac{1}{2\lambda^2}$, there are a region of unbounded motion $y > C/\lambda$, and the region with bounded motion

$$y_3 = \frac{C}{2} \le y(x) < \frac{C}{\lambda} \qquad (6.42)$$

which can be analogous to a glacial valley profile. In the limit $\lambda \to 1^+$ one of the turning points (6.39) moves to infinity and disappears, leaving the single turning point $y_3 = C/2$.

6.3.1.3   *Parameter range $0 < \lambda < 1$*

Consider now values $\lambda < 1$ of the Lagrange multiplier, corresponding to

$$-\frac{1}{2\lambda^2} < E = -\frac{1}{2} ; \qquad (6.43)$$

now the horizontal line $E = -1/2$ is above the horizontal asymptote of $V(y)$ and intersects the graph of $V(y)$ only once for positive y. There is a region of bounded motion

$$y_4 \le y(x) < \frac{C}{\lambda} \qquad (6.44)$$

and this situation is qualitatively similar to the case $\lambda = 1$. The turning point $y_4 = C/(1 - \lambda)$ is in the physical region $y > 0$ and the second turning point $y_5 = -C/(1 + \lambda) < 0$ is unphysical. The analytical solution (6.23) of the Morgan equation (6.17) [86] belongs to this parameter range.

6.3.1.4 *Negative parameter* λ

It is time to discuss the second condition $C > 0$ of Ref. [286] that we have assumed to hold. Due to the symmetry of Eq. (6.17), the region of parameter space $C < 0$ and $\lambda > 0$ is equivalent to the region $C > 0$ and $\lambda < 0$ discussed below.

The vertical asymptote $y = C/\lambda$ of the potential energy $V(y)$ lies in the unphysical $y < 0$ region. The derivative $V'(y)$ is negative for $y < C/\lambda < 0$ and for $y > 0$, and is positive for $C/\lambda < y < 0$. If the Lagrange multiplier has values $-1 < \lambda < 0$, corresponding to $E = -\dfrac{1}{2} > -\dfrac{1}{2\lambda^2}$, there is only one intersection $y_6 = C/(1+\lambda)$ in the $y > 0$ region and the particle motion is unbounded.

The only remaining possibility is $C = 0$, for which Eq. (6.17) reduces to $(y')^2 = -1 + 1/\lambda^2$. Only linear solutions exist, which could be used to model V–shaped valleys. These are not relevant as glacial valley cross–profiles but sometimes are used as initial conditions in numerical simulations [370]. The assumption $C > 0$ is then justified.

6.3.1.5 *Analogue of the unconstrained variational problem*

Using the particle analogy, the incorrect Eq. (6.30) corresponding to extremization without constraints maps into the energy integral

$$\frac{(\xi')^2}{2} + U(\xi) = E \,, \tag{6.45}$$

where $E = -1/2$ and the potential energy

$$U(\xi) = -\frac{\xi^2}{2C_2^2} \tag{6.46}$$

corresponds to an inverted harmonic oscillator. Apart from the unstable equilibrium position $\xi \equiv 0$, which is meaningless as a valley profile, all motions are unbounded and unphysical in the glaciology problem.

6.3.2 *Cosmological analogy*

Let us discuss now the analogy between the Morgan and the Friedmann equations [141, 154]. The most straightforward identification between Eq. (6.17) and the Friedmann equation (1.179) is obtained by setting $\kappa = 1$, giving the analogous Friedmann equation

$$\frac{\dot{a}^2}{a^2} = \frac{1}{(\lambda a - C)^2} - \frac{1}{a^2} \tag{6.47}$$

or

$$\frac{\dot{a}^2}{a^2} = \frac{8\pi G}{3} \frac{\rho_0}{(a - a_0)^2} - \frac{1}{a^2}, \tag{6.48}$$

where

$$\rho_0 = \frac{3}{8\pi G \lambda^2}, \qquad a_0 = \frac{C}{\lambda} \tag{6.49}$$

are positive constants. The energy density of the cosmic fluid

$$\rho(t) = \frac{\rho_0}{(a - a_0)^2} \tag{6.50}$$

is positive and diverges when $a \to a_0$, which corresponds to a spacetime singularity. The fluid pressure follows by imposing the conservation equation (1.187),

$$P = \frac{2\rho_0 \, a}{3 \, (a - a_0)^3} - \frac{\rho_0}{(a - a_0)^2}. \tag{6.51}$$

An alternative expression is

$$P = -\frac{\rho}{3} \pm \frac{2a_0}{3\sqrt{\rho_0}} \rho^{3/2}, \tag{6.52}$$

where the upper sign applies when $a > a_0$ and the lower sign when $a < a_0$. The condition $\rho + P > 0$ corresponds to $a > a_0$. Cosmic equations of state corresponding to the lower sign in (6.52) and violating the weak energy condition are discussed in [26]. More complicated equations of state $P = \sum\limits_{k=1}^{m} c_k \, \rho_{(k)}^k$ are discussed in [86, 87] and quadratic ones in [6,7,78,297,298,372]. All these exotic equations of state, including (6.52), originate unusual cosmological singularities.

The acceleration equation (1.180) reads

$$\frac{\ddot{a}}{a} = -\frac{a_0}{\lambda^2 (a - a_0)^3}. \tag{6.53}$$

The analogous cosmos accelerates if $a < a_0$ and decelerates if $a > a_0$. The value a_0 of the scale factor identifies a spacetime singularity since the Ricci scalar

$$\mathcal{R} = 8\pi G \, (\rho - 3P) = \frac{16\pi G \rho_0}{(a - a_0)^3} (a - 2a_0) \tag{6.54}$$

diverges as $a \to a_0$. This value of the scale factor is actually approached during the dynamics. The Friedmann equation (1.179) amounts to

$$a^2 \geq \lambda^2 (a - a_0)^2 \tag{6.55}$$

and excludes the orbits of the solutions of the dynamical system (1.179), (1.187) from a region of the $\left(a, \dot{a} \right)$ phase space.

- If $a > a_0$, the constraint (6.55) is written as $a \leq \lambda a_0/(\lambda - 1)$. The coefficient of a_0 is

$$\frac{\lambda}{\lambda - 1} = 1 + \frac{1}{\lambda - 1} > 1 \qquad (6.56)$$

and in this regime it is

$$a_0 < a < \frac{\lambda a_0}{\lambda - 1}. \qquad (6.57)$$

The scale factor $a(t)$ is bounded from above but nothing forbids it from approaching the value a_0 arbitrarily close.

- In the other case $a < a_0$ the constraint (6.55) becomes

$$a \geq \frac{\lambda a_0}{\lambda + 1} \qquad (6.58)$$

and

$$0 < \frac{\lambda a_0}{\lambda + 1} \leq a < a_0. \qquad (6.59)$$

In both cases the scale factor is bounded from below by a positive constant and cannot have Big Bang or Big Crunch singularities $a \to 0$ [82, 260, 435] but it can reach the singularity a_0. The fact that one cannot have $y = 0$ in the glaciology side of the analogy because it implies an imaginary y', noted as "a curious feature" in Ref. [286], is more transparent now: the value $y = 0$ (or $a = 0$) lies in the region of the phase space forbidden by the constraint (6.55). The latter corresponds to the general Hamiltonian constraint of general relativity, *i.e.*, the first order time–time component of the Einstein equations [435].

The boundaries $a = \lambda a_0/(\lambda \mp 1)$ are formal solutions of Eq. (6.47) with constant scale factor, but they violate the acceleration equation (1.180) and are meaningless as glacial valley profiles.

The acceleration equation shows that, if $a > a_0$, then $\ddot{a} < 0$, the function $a(t)$ has concavity facing downwards, and the cosmos decelerates. $a(t)$ is continuous, then it always decreases until it crosses the horizontal line $a = a_0$ (Fig. 6.4).

Using Eq. (6.49), we now recast Eq. (6.47) in the asymptotic form

$$\dot{a}^2 = \frac{a^2}{\lambda^2 (a - a_0)^2} - 1 \approx \frac{a_0^2}{\lambda^2 (a - a_0)^2} \qquad (6.60)$$

as $a \to a_0$, which has solution

$$a(t) \simeq a_0 \pm \sqrt{\frac{2a_0}{\lambda}} |t - t_0|, \qquad (6.61)$$

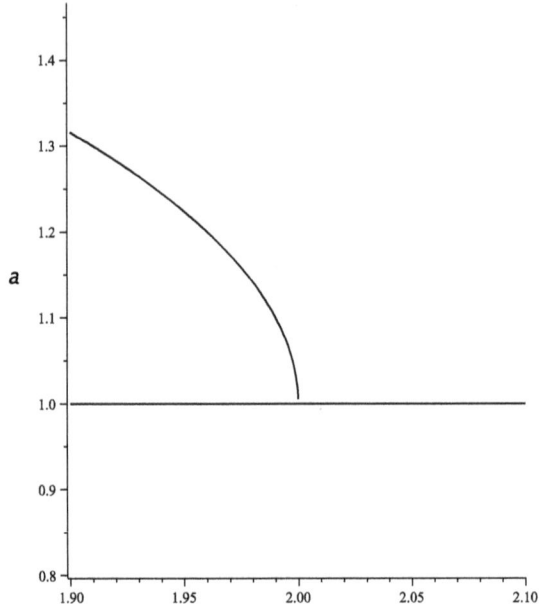

Legend

———— scale factor $a(t)$

FIG. 6.4. The behaviour of the scale factor $a(t)$ near the singularity a_0 when $a > a_0$, as described by Eq. (6.61) with positive sign (for the parameter values $a_0 = 1$, $t_0 = 2$). The concavity faces downward and the universe contracts.

where we choose the positive sign in front of the square root because $a > a_0$. The integration constant t_0 is the time at which the singularity occurs. This solution is the analogue of a meaningful glacial valley profile because y is the ice thickness, which is maximum at $x = 0$ and minimum at the boundaries [286]. It is also an interesting solution in FLRW cosmology because it shows an example of a finite time singularity without violating the weak energy condition [82,435] $\rho > 0$ and $\rho + P > 0$, a scenario discussed in Ref. [26].

If, instead, $a < a_0$, the universe accelerates with $\ddot{a} > 0$ and the scale factor $a(t)$ always increases until it crosses the horizontal line $a = a_0$ (Fig. 6.5). The solution of the asymptotic equation (6.60) is

$$a(t) \simeq a_0 - \sqrt{\frac{2a_0}{\lambda}} \, |t - t_0| , \qquad (6.62)$$

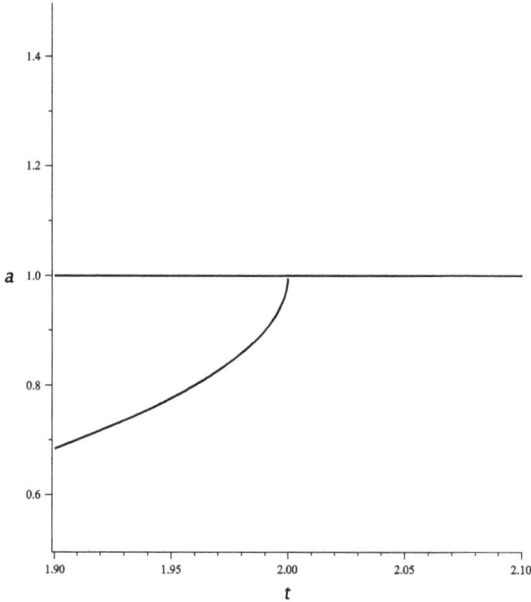

Legend

——— scale factor $a(t)$

FIG. 6.5. The behaviour of the scale factor $a(t)$ near the singularity a_0 when $a < a_0$, as described by Eq. (6.62) (for the parameter values $a_0 = 1$, $t_0 = 2$). The concavity faces upward and the expanding universe is accelerated.

where we choose the negative sign because $a < a_0$. This time, there is no glacial valley profile on the other side of the analogy. The slope of $a(t)$ diverges where $a(t)$ crosses the line $a = a_0$, singularity of the analogous cosmos, which reaches the minimum size

$$a_{\min} = \frac{\lambda a_0}{\lambda + 1} \tag{6.63}$$

and then bounces. This behaviour requires the violation of the weak energy condition, which happens because

$$\rho + P = \frac{2\rho_0 a}{3(a - a_0)^3} < 0. \tag{6.64}$$

The cosmological singularity is not the Big Rip occurring with phantom energy [72, 73]: here the scale factor $a(t)$ remains finite but the Hubble function $H(t)$, the energy density ρ, the pressure P, and the Ricci scalar $\mathcal{R}$

all diverge as $a \to a_0$. This is the spatially curved counterpart of the Big Freeze [56] or Type III [16] singularity studied in spatially flat universes. A Big Freeze is known to occur also in alternative theories of gravity, such as Palatini $f(\mathcal{R})$ gravity (*e.g.*, [398]).

Near the Big Freeze the main features remain the same as in $\kappa = 0$ universes [141, 154] because, as $a \to a_0$ in Eq. (6.48), the energy density is dominated by the divergent term proportional to $(a - a_0)^{-2}$, which shadows the finite curvature term $-1/a^2$. Finite time singularities are the subject of an extensive literature in FLRW cosmology [16, 26, 29, 36, 56, 70, 72, 73, 98, 99, 163, 185, 243, 356].

For completeness, we mention also the cosmic analogue of Eq. (6.17) corresponding to $C = 0$ (noted in Ref. [87]), $a_0 = 0$, and to

$$H^2 = \frac{1}{a^2} \left(\frac{1 - \lambda^2}{\lambda^2} \right) , \qquad (6.65)$$

which implies $|\lambda| \leq 1$. The counterpart on the glaciology side of the analogy is unphysical: $\dot{a} = $ const. and the solutions are linear in time, containing the static universe with $a = $ const. as a special case.

Our discussion in this chapter is not exhaustive and other aspects deserve further investigation. To begin with, the friction of glacier ice against the valley walls and bed was described by the Coulomb law model, but it could include viscous friction which depends on the velocity. Indeed, a friction model quadratic in the velocity is used in numerical simulations of the dynamical erosion process [370]. Then, the numerical studies of the formation process of glacial valleys ignore the variational approach, although they include its Lagrangian constraint, but no comparison between the two approaches exist in the literature.

6.3.3   *Maximum or minimum friction?*

As we have seen, there was a debate concerning the question of whether friction in the glacial valley model is maximized or minimized. It seems logical that it should reach a maximum but the justification of this expectation is not trivial. Invoking the second law of thermodynamics is not sufficient: this law states that the entropy of an isolated system (or of the thermodynamical universe[2] consisting of an open system plus its surroundings) never decreases [83]. However, entropy and friction do not coincide and the frictional system considered here is not isolated, since it is driven by climate

---

[2]This word is used here in its thermodynamical sense [83], which has nothing to do with cosmology.

forcing from the atmosphere. Furthermore, energy is exchanged as heat between the glacier and the valley walls, through subglacial hydrology and other processes. Driven systems are not isolated but have energy sources. Even worse, although only the final state is described by the variational principle used, the dissipative process involves non–equilibrium thermodynamics.

The original work of Hirano & Aniya [212] stated that friction is minimized in the glacial erosive process. This viewpoint was criticized in Ref. [197] (see also [213, 214, 286]) arguing that friction should be maximized instead. Although this issue is immaterial for the first order variational principle $\delta J\left[y(x)\right] = 0$ leading to the final glacial valley profiles, it is important to understand the physics. This problem is analogous to the one encountered for equilibrium beach profiles and discussed in Chap. 5, which reserved a surprise when investigated and is reminiscent of similar problems in the earth sciences. Here, intuition suggests that friction is maximized instead of minimized but, in earth science problems involving geophysical flows, the opposite point of view is often supported. The rationale is that minimizing friction optimizes the flow, as seen in Chap. 5 for equilibrium beach profiles.

The issue for glacial valley profiles is settled in Ref. [148]: the final profile maximizes friction. To reach this conclusion it is again sufficient to trust the mathematics of the glacial valley profile model already used and to compute the second variation of the functional $J\left[y(x)\right]$, assessing its sign. The structure of the calculation parallels the one already seen for equilibrium beach profiles but it differs in the details, of course, because the Lagrangians are different.

Consider again the action functional for the glacial valley problem

$$J = \int_{-x_0}^{+x_0} dx \left(y\sqrt{1 + y'^2} - \lambda y \right) \equiv \int_{-x_0}^{+x_0} dx\, L\left(y, y'\right) \qquad (6.66)$$

and the regular variations $\eta(x)$ around the actual extremizing path that vanish at the endpoints, $\eta(-x_0) = \eta(x_0) = 0$, which are parametrized by the parameter α. The varied paths are [438]

$$y(x, \alpha) = y(x, 0) + \alpha\, \eta(x), \qquad (6.67)$$

where $y(x, 0)$ is the extremizing glacial valley profile labelled by $\alpha = 0$. We need the properties

$$y'(x, \alpha) = y'(x, 0) + \alpha \frac{d\eta}{dx}, \tag{6.68}$$

$$\frac{\partial y}{\partial \alpha} = \eta, \tag{6.69}$$

$$\frac{\partial y'}{\partial \alpha} = \frac{d\eta}{dx}, \tag{6.70}$$

$$\frac{\partial^2 y'}{\partial \alpha^2} = \frac{\partial^2 y}{\partial \alpha^2} = 0, \tag{6.71}$$

that lead to the second variation of the action [438]

$$\frac{\partial^2 J}{\partial \alpha^2} = \int_{-x_0}^{+x_0} dx \left[\frac{\partial^2 L}{\partial y'^2} \left(\frac{d\eta}{dx} \right)^2 + 2 \frac{\partial^2 L}{\partial y \, \partial y'} \eta \frac{d\eta}{dx} + \frac{\partial^2 L}{\partial y^2} \eta^2 \right]. \tag{6.72}$$

Computing the derivatives

$$\frac{\partial L}{\partial y} = \sqrt{1 + y'^2} - \lambda, \tag{6.73}$$

$$\frac{\partial^2 L}{\partial y^2} = 0, \tag{6.74}$$

$$\frac{\partial^2 L}{\partial y \, \partial y'} = \frac{y'}{\sqrt{1 + y'^2}}, \tag{6.75}$$

$$\frac{\partial L}{\partial y'} = \frac{y \, y'}{\sqrt{1 + y'^2}}, \tag{6.76}$$

$$\frac{\partial^2 L}{\partial y'^2} = \frac{y \left[y'' \left(1 + y'^2 \right) - y'^2 \right]}{\left(1 + y'^2 \right)^{3/2}}, \tag{6.77}$$

and substituting them into the integral (6.72) yields

$$\frac{\partial^2 J}{\partial \alpha^2} = \int_{-x_0}^{+x_0} dx \left\{ \frac{y \left[y'' \left(1 + y'^2 \right) - y'^2 \right]}{\left(1 + y'^2 \right)^{3/2}} \left(\frac{d\eta}{dx} \right)^2 + \frac{y'}{\sqrt{1 + y'^2}} \frac{d(\eta^2)}{dx} \right\}, \tag{6.78}$$

where the integral is computed along the extremizing trajectories for which $\partial J / \partial \alpha = 0$.

We use the known facts that $y(x) \geq 0 \ \forall x \in \left[-x_0, x_0 \right]$ and y vanishes only at the boundaries $x = \pm x_0$, and that $y''(x) < 0 \ \forall x$. Then, it must be

$$y y'' \left(1 + y'^2 \right) < 0 \qquad (6.79)$$

for $x \in \left(-x_0, x_0 \right)$. With the help of the information that

$$- y(x) y'^2(x) \leq 0 \qquad \forall x, \qquad (6.80)$$

we reach the conclusion that

$$\frac{y \left[y'' \left(1 + y'^2 \right) - y'^2 \right]}{\left(1 + y'^2 \right)^{3/2}} \left(\frac{d\eta}{dx} \right)^2 \leq 0 \qquad \forall x \in \left(-x_0, x_0 \right) \qquad (6.81)$$

and the first integral in Eq. (6.78) is non–positive,

$$\int_{-x_0}^{+x_0} dx \ \frac{y \left[y'' \left(1 + y'^2 \right) - y'^2 \right]}{\left(1 + y'^2 \right)^{3/2}} \left(\frac{d\eta}{dx} \right)^2 \leq 0. \qquad (6.82)$$

Passing to the second integral and integrating by parts, we have

$$\int_{-x_0}^{+x_0} dx \ \frac{y'}{\sqrt{1 + y'^2}} \frac{d(\eta^2)}{dx} = \left. \frac{\eta^2 \, y'}{\sqrt{1 + y'^2}} \right|_{-x_0}^{+x_0} - \int_{-x_0}^{+x_0} dx \, \eta^2 \frac{d}{dx} \left(\frac{y'}{\sqrt{1 + y'^2}} \right). \qquad (6.83)$$

The first term on the right hand side is identically zero due to the fact that the variation $\eta(x)$ vanishes at the endpoints $x = \pm x_0$. Then, we use the fact that the extremizing trajectories $y(x)$ obey the Morgan equation with positive constant C. The substitution of this equation into the integral (6.83) gives

$$I \equiv - \int_{-x_0}^{+x_0} dx \, \eta^2 \frac{d}{dx} \left(\frac{y'}{\sqrt{1 + y'^2}} \right)$$

$$= - \int_{-x_0}^{+x_0} dx \, \eta^2 \frac{d}{dx} \left[\left(\lambda - \frac{C}{y} \right) y' \right]$$

$$= - \int_{-x_0}^{+x_0} dx \, \eta^2 \frac{C \, y'^2}{y^2} + \int_{-x_0}^{+x_0} dx \left(\frac{C}{y} - \lambda \right) y'' \qquad (6.84)$$

$$\equiv I_1 + I_2. \qquad (6.85)$$

We now notice that the integrand of the first integral I_1 is non–negative and $I_1 \leq 0$. Proceeding with the second integral

$$I_2 = \int_{-x_0}^{0} dx \left(\frac{C}{y} - \lambda \right) y'' + \int_{0}^{+x_0} dx \left(\frac{C}{y} - \lambda \right) y'', \qquad (6.86)$$

note that $C/y \to +\infty$ at the valley boundaries $\pm x_0$ with $y(x) \to 0^+$; then, both integrals on the right hand side of Eq. (6.86) are dominated by the regions around $\pm x_0$, where $\dfrac{C}{y} - \lambda \simeq \dfrac{C}{y}$ and λ is finite. Due to the fact that $y''(x) < 0 \ \forall x \in \left(-x_0, x_0 \right)$, both integrals are negative and one concludes that $I_2 < 0$.

We can finally combine these partial results to write

$$\frac{\partial^2 J}{\partial \alpha^2} = \underbrace{\int_{-x_0}^{+x_0} dx \, \frac{y\left[y'' \left(1 + y'^2\right) - y'^2\right]}{\left(1 + y'^2\right)^{3/2}} \left(\frac{d\eta}{dx}\right)^2}_{< 0}$$

$$\underbrace{- \int_{-x_0}^{+x_0} dx \, \frac{C \, y'^2 \, \eta^2}{y^2}}_{< 0} + \underbrace{\int_{-x_0}^{+x_0} dx \left(\frac{C}{y} - \lambda\right) y''}_{< 0} \qquad (6.87)$$

and we can state the final result as

$$\frac{\partial^2 J}{\partial \alpha^2} < 0 : \qquad\qquad (6.88)$$

the equilibrium valley profiles that extremize the action J maximize the friction [148].

6.4   The universe in a glacier slope

The longitudinal profile of glaciers and ice caps is an important geomorphological feature studied in glaciology (Fig. 6.6).

The longitudinal profile is the ice thickness $h(x)$ as a function of a coordinate x, with the corresponding x–axis pointing downstream along the glacier bed, which is assumed to be a plane. In models of glaciers and ice caps, the longitudinal profile is not arbitrary but is determined by the physical assumptions used to build the model. Moreover, measuring the ice thickness, the longitudinal profile enters the estimate of the volume and mass of ice contained in a glacier or ice sheet, or in a certain region [139]. For glaciers and ice caps, the volume scales with the surface area in a well–defined way and this volume–area scaling is a major tool used in the estimate of the ice content, ice loss, and sea level rise related to climate change [15]. In fact, while it is relatively easy to estimate the surface area from aerial photographs, measuring the thickness to estimate the volume is much more difficult, so the volume–area scaling law is invaluable in this context.

Glacier models are relatively well developed and make definite predictions about longitudinal glacier profiles, which depend crucially on the rheology of glacier ice. Let s_{ij} be the deviatoric stresses in the ice and

$$\sigma_{\text{eff}} = \sqrt{\frac{1}{2}\,\text{Tr}\left(\hat{s}^2\right)} \tag{6.89}$$

be the effective stress [97, 216, 323]. The ice rehology is described by Glen's law relating the response of glacier ice (*i.e.*, the strain rate tensor $\dot{\epsilon}_{ij}$) with the applied deviatoric stresses s_{ij}. Glen's law is [182]

$$\dot{\epsilon}_{ij} = \mathcal{A}\,\sigma_{\text{eff}}^{n-1}\,s_{ij}\,, \tag{6.90}$$

where $\mathcal{A}$ is a (temperature–dependent) constant [97, 186, 216, 323]. The strain rate $\dot{\epsilon}_{ij}$ depends on all components of the deviatoric stresses (*i.e.*, in all directions) through its dependence on σ_{eff}. Various values of the exponent n describe very different physics: at very low stresses, $n = 1$ corresponds to a perfectly viscous material with viscosity coefficient $\eta = \mathcal{A}^{-1}$, which obeys the linear relation $s_{ij} = \dot{\epsilon}_{ij}/\mathcal{A}$ [97, 216, 323]. The different

FIG. 6.6. Glaciers below Mt. Angelo Grande/Hoher Angelus (3521 m) in the Ortler Alps of South Tyrol, Italy.

value $n = 2$ is used to model the sliding of a glacier along its bed due to lubrication and till deformation [97, 323]. The value $n = 3$, in which we are interested here, describes ice flow due to its creep and is the value relevant for modelling ice flow in the vast majority of glaciers [97, 186, 216, 323]. Finally, the limit $n \to +\infty$ corresponds to the approximation of perfectly plastic ice, which is a poor approximation to real ice flow [97, 186, 216, 323].

Glacier and ice cap models assume incompressible and isotropic ice, steady state, a flat bed (a plane with non–vanishing slope, although polar ice caps are usually taken to have horizontal bed in these models), the shallow ice approximation (thickness much smaller than length), and Glen's law to describe ice creep [97, 186, 216, 323]. These simplifications are relaxed in the numerical modelling of glaciers, but are necessary for an analytical description.

Let ρ_{ice} be the ice density, g the acceleration of gravity, and $c(x)$ the accumulation rate of ice. The latter is the flux density of ice volume in the z–direction, which is taken to be perpendicular to the x–axis and to the flat bed. c has the dimensions of a velocity.

Under the assumptions above, it is found [97, 186, 216, 323, 418] that the longitudinal glacier profile $h(x)$ of a glacier satisfies the Vialov ordinary differential equation

$$x\, c(x) = \frac{2A}{n+2}\left(\rho_{\mathrm{ice}}\, g\, h\left|\frac{dh}{dx}\right|\right)^{n} h^{2},\tag{6.91}$$

which expresses the mass conservation of ice.

Several aspects of glaciology require analytical expressions for longitudinal glacier profiles (*e.g.*, [15, 37, 139, 294, 403]). Solutions with finite length L and $x \in \left[0, L\right]$ usually locate the glacier summit at $x = 0$ and its terminus at $x = L$. Then, the slope dh/dx is negative and its absolute value is used in the Vialov equation. It is, of course, possible to switch the locations of the summit and the terminus, and then the slope dh/dx becomes positive in $\left(0, L\right)$. The use of the absolute value in the Vialov equation (6.91) covers both situations.

Once the solution for the longitudinal profile of half of a glacier is determined in the interval $\left[0, L\right]$, it is reflected about the vertical line $x = 0$ (or $x = L$, respectively) passing through the summit and extended by continuity to the interval $\left[-L, L\right]$ (or to $\left[0, 2L\right]$, respectively). As a consequence, the ice thickness profile $h(x)$ of an ice cap or ice sheet is continuous but not differentiable at the summit because left and right derivatives h' will have opposite signs at this point.

At the glacier terminus, both the slope dh/dx and the basal stress

$$\tau_b = -\rho_{\text{ice}} \, g \, h \, \frac{dh}{dx} \tag{6.92}$$

diverge. This unphysical feature is a well known shortcoming of the models associated with the fact that the shallow ice approximation used in the derivation of the Vialov equation breaks down at the terminus [97, 186, 216, 323].

The Vialov equation (6.91) is formally integrated as

$$h(x) = \left\{ \mp \frac{2\,(n+1)}{n\,\rho_{\text{ice}}\,g} \left(\frac{n+2}{2\mathcal{A}} \right)^{1/n} \int dx \Big[x\,c(x) \Big]^{1/n} \right\}^{\frac{n}{2(n+1)}}$$

$$\equiv A_0 \Big[I(x) \Big]^{\frac{n}{2(n+1)}}, \tag{6.93}$$

where the upper sign applies if the summit is at $x = 0$ and $dh/dx < 0$, while the lower sign describes the situation in which $x = L$ is the summit. The constant A_0 and the integral I are given by

$$A_0 \equiv \left[\frac{2\,(n+1)}{n\,\rho_{\text{ice}}g} \left(\frac{n+2}{2\mathcal{A}} \right)^{1/n} \right]^{\frac{n}{2(n+1)}}, \tag{6.94}$$

and

$$I(x) \equiv \int dx \Big[x\,c(x) \Big]^{1/n} \tag{6.95}$$

(I is defined up to an integration constant). The model requires the specification of the function $c(x)$ describing the accumulation rate of ice. Ideally, the choice of $c(x)$ is motivated by atmospheric models of the precipitation in the region considered. Unfortunately, even with simple choices of the function $c(x)$ it is impossible to compute the integral (6.95) explicitly in terms of elementary functions. A few analytical solutions of the Vialov equation are, however, known [46, 67, 68, 322, 418, 439].

The Chebyshev theorem of integration used to integrate the Friedmann equation can be used to characterize the models that admit explicit elementary solutions [140]. There are also analytical longitudinal profiles obtained under the unrealistic assumption of perfectly plastic ice. This assumption was used in the early days of glacier modelling, but it appropriate only when the deformation of the ice is negligible [143, 155, 302, 303]. It corresponds to the limit $n \to +\infty$ of the Vialov equation [97, 186, 216, 323].

We will show below that, with a suitable rescaling of the independent variable x, the Vialov equation admits Lagrangian and Hamiltonian formulations, solving its inverse variational problem [149]. However, in glacier models the Vialov equation is not obtained from a variational principle. In general, the inverse variational problem of mathematical physics of finding Lagrangians and Hamiltonians for a given differential equation (Helmoltz problem [209]), can be solved in a surprisingly wide number of cases [266]. One possible approach uses non–standard Lagrangians for dissipative–like autonomous systems of differential equations [89,290,291,355]. We will not use this technique here, but our procedure uses a redefinition of variables similar to some of the equations used in [355]. The Lagrangian formulation and the conservation of the corresponding Hamiltonian reproduce the Vialov equation and lay the basis for another cosmological analogy. As shown below, this analogy gives back to glaciology by allowing, at least formally, the generation of all solutions of the Vialov equation from a generating function, which is the Nye parabolic profile for perfectly plastic ice.

6.4.1   *Lagrangian formulation of the Vialov equation*

In order to solve the inverse variational problem, we recast the Vialov equation (6.91) as

$$h^{\frac{n+2}{n}} h' = \left[\frac{x\, c(x)}{\alpha} \right]^{1/n} , \qquad (6.96)$$

where

$$\alpha = \frac{2\mathcal{A}}{n+2} \left(\rho_{\text{ice}}\, g \right)^n . \qquad (6.97)$$

We then change the independent variable according to $x \to \bar{x}$ with

$$d\bar{x} = \left[x\, c(x) \right]^{1/n} dx \qquad (6.98)$$

or, in finite form,

$$\bar{x}(x) = \int_0^x dx' \left[x'c(x') \right]^{1/n} . \qquad (6.99)$$

This change of variable is invertible and one–to–one where $c(x) > 0$, then $d\bar{x}/dx > 0$ and we require $c(x)$ to be continuous and piecewise differentiable.[3]

---

[3]One exception is of interest in glaciology, the Weertman–Paterson model [322, 439] in which $n = 3$ and $c(x)$ is given as a Heaviside step function on the adiacent intervals $(0, R)$ and (R, L), on which it is piecewise constant and has opposite sign. This situation is reduced to the previous one by studying the two intervals separately and matching the corresponding solutions at $x = R$.

Using $\bar{x}$, the Vialov equation becomes

$$h^{\frac{n+2}{n}}\left|\frac{dh}{d\bar{x}}\right| = \frac{1}{\alpha^{1/n}}.$$ (6.100)

With the now familiar procedure of dividing by $\sqrt{2}\,h^{\frac{n+2}{n}}$ and squaring, we obtain the Friedmann–like equation

$$\left(\frac{dh}{d\bar{x}}\right)^2 = \frac{1}{\alpha^{2/n}}\,h^{-\frac{2(n+2)}{n}},$$ (6.101)

which can also be interpreted as the energy integral for a particle of unit mass at position $h(\bar{x})$ at time $\bar{x}$ under the conservative force $-dV/dh$, where

$$V(h) = -\frac{V_0}{h^{\frac{2(n+2)}{n}}}$$ (6.102)

is the potential energy and

$$V_0 = \frac{1}{2\,\alpha^{2/n}}.$$ (6.103)

The corresponding Lagrangian is obviously

$$\mathcal{L}\left(h(\bar{x}), \frac{dh}{d\bar{x}}(\bar{x})\right) = \frac{1}{2}\left(\frac{dh}{d\bar{x}}\right)^2 - \frac{V_0}{h^{\frac{2(n+2)}{n}}}.$$ (6.104)

The scenario degenerates in the limit $n \to +\infty$ of perfectly plastic ice, in which $\bar{x}$ coincides with x, the accumulation function $c(x)$ disappears, and the Lagrangian (6.104) degenerates into

$$\mathcal{L}_\infty\left(h(\bar{x}), \frac{dh}{d\bar{x}}(\bar{x})\right) = \frac{1}{2}\left(\frac{dh}{d\bar{x}}\right)^2 - \frac{V_0}{h^2}.$$ (6.105)

When n is finite, the Lagrangian (6.104) is independent of $\bar{x}$, which implies the conservation of the associated Hamiltonian

$$\mathcal{H} = \frac{\partial\mathcal{L}}{\partial(dh/d\bar{x})}\frac{dh}{d\bar{x}} - \mathcal{L} = \frac{1}{2}\left(\frac{dh}{d\bar{x}}\right)^2 - \frac{1}{2\,\alpha^{2/n}\,h^{\frac{2(n+2)}{n}}} = E,$$ (6.106)

where E is the constant energy of the analogous particle. As usual, the Vialov equation is only obtained when this Hamiltonian vanishes (corresponding to the Hamiltonian constraint of general relativity [82, 435]).

Assuming that we have found an analytical solution $h(\bar{x})$ of Eq. (6.106), we then need to replace $\bar{x}$ with x. It may be impossible to complete this step explicitly using only elementary functions. Again, the Chebyshev theorem of integration characterizes the situations when the integral $I(x)$ involves only elementary functions [140].

Let us consider, as an example, the original model by Vialov [418] corresponding to the choice of accumulation function $c(x) = c_0 \equiv \text{const.}$, which yields

$$\bar{x}(x) = \frac{n\, c_0^{1/n}}{n+1}\, x^{\frac{n+1}{n}} . \qquad (6.107)$$

We look for a solution defined on $x \in \left[0, L\right]$, locating the glacier summit at $x = \bar{x} = 0$ where $h(0) = H$ and its terminus at $x = L$ where $h = 0$ (and, necessarily, $dh/dx \to \infty$). Then the relevant differential equation reads

$$h^{\frac{n+2}{n}} \frac{dh}{d\bar{x}} = -\frac{1}{\alpha^{1/n}}, \qquad (6.108)$$

and its solution is

$$\left[h(\bar{x})\right]^{\frac{2(n+1)}{n}} = H^{\frac{2(n+1)}{n}} \left[1 - \frac{2(n+1)\bar{x}}{n\alpha^{1/n} H^{\frac{2(n+1)}{n}}}\right]. \qquad (6.109)$$

By raising both sides to the power $\dfrac{n}{2(n+1)}$, we have

$$h(\bar{x}) = H \left[1 - \frac{2\,(n+1)\,\bar{x}}{n\,\alpha^{1/n} H^{\frac{2(n+1)}{n}}}\right]^{\frac{n}{2(n+1)}} ; \qquad (6.110)$$

Equation (6.107) then allows us to express this solution in terms of the original coordinate x along the glacier bed [149]

$$h(x) = H \left[1 - \left(\frac{x}{L}\right)^{\frac{n+1}{n}}\right]^{\frac{n}{2(n+1)}}, \qquad (6.111)$$

with

$$L = \frac{1}{2^{\frac{n}{n+1}}} H^2 \left(\frac{\alpha}{c_0}\right)^{\frac{1}{n+1}} : \qquad (6.112)$$

this is the classic Vialov profile [418].

6.4.2   *Point–particle analogy*

Let us discuss the mechanical analogy that we used to solve the inverse variational problem for the Vialov equation. Equation (6.106) looks like the energy integral for a particle of unit mass and total mechanical energy E in the potential energy, with the actual physical situation corresponding only to the energy value $E = 0$. Only the interval $h \in \left[0, H\right]$ is physical. Since $n > 0$ for all practical applications, with $n = 3$ for ice creep, the potential $V(h)$ is negative with $V(h) \leq V(H) < 0$, is concave downward, and has a vertical asymptote, where $V(h) \to -\infty$ as $h \to 0^+$.

The range of the particle with total energy $E = 0$ is truncated artificially at the glacier summit where $h = H$, where $h(x)$ is reflected about the vertical line through the summit, which means that in the (h, V) plane the potential energy $V(h)$ in $h \in \left(H, 2H\right)$ is replaced with its reflection in $\left(0, H\right)$ about the line $h = H$ and the x–axis with $h > 2H$ is irrelevant.

The artificial truncation and the confinement to $h \in \left(0, L\right)$ deform the scenario of the possible motions obtained with the Weierstrass method employed in similar situations previously encountered in this book. We see that, for the particle to maintain zero mechanical energy, the kinetic energy diverges as the particle falls into the infinite potential well at $h = 0$ (analogous to the glacier terminus). Then, the derivative dh/dx must diverge there, which mirrors the unphysical divergence of the basal stress $\tau_b = -\rho_{\text{ice}}\, gh\, dh/dx$ at the glacier terminus in the Vialov equation, as already discussed.

6.4.3 *Cosmological analogy*

Let us discuss the (now obvious) analogy between the reduced Vialov equation and the Friedmann equation. We rewrite Eq. (6.106) in the Friedmann–like form

$$\left(\frac{1}{h}\frac{dh}{d\bar{x}}\right)^2 = \frac{1}{2\,\alpha^{2/n}\,h^{\frac{4(n+1)}{n}}} + \frac{E}{h^2}. \qquad (6.113)$$

The analogous universe has scale factor $a(t)$ which mirrors $h(\bar{x})$, is spatially flat with curvature index $\kappa = -E = 0$, and is filled with a perfect fluid with equation of state parameter

$$w = \frac{n+4}{3n} \qquad (6.114)$$

and positive–definite energy density

$$\rho(a) = \frac{\rho_0}{a^{3(w+1)}} = \frac{\rho_0}{a^{4(n+1)/n}}, \qquad (6.115)$$

$$\rho_0 = \frac{3}{16\pi G\,\alpha^{2/n}}. \qquad (6.116)$$

The parameter value $n = 3$ describing ice creep [182] corresponds to the cosmic equation of state parameter $w = 7/9 \simeq 0.778$, which makes the analogous universe decelerate. In general, the equation of state parameter $w > 0$

is a monotonically decreasing function of n which diverges as $n \to 0$ and has a horizontal asymptote $w = 1/3$ (describing a radiation era) of the scale factor which, in the analogy, mirrors the perfectly plastic ice limit $n \to +\infty$. The Hubble function H diverges when $a \to 0$, which is either a Big Bang or a Big Crunch.

Remember that the vertical slope of the glacier $h' \to -\infty$ at its terminus $x = L$ is associated with the breakdown of the shallow ice approximation in the Vialov equation. On the cosmological side of the analogy, the shallow ice approximation translates into small $\dot{a}$ or, roughly speaking, $a(t)/t \ll 1$. Because $a(t) = a_0\, t^{\frac{2}{3(w+1)}}$, the ratio

$$\frac{a(t)}{t} = \frac{a_0}{t^{\frac{3w+1}{3(w+1)}}} \tag{6.117}$$

diverges at the Big Bang or the Big Crunch. This occurrence prompts a reflection: it is expected that quantum mechanics will change the behaviour of matter and of gravity and will make the Hawking–Penrose singularity theorems inapplicable, preventing the singularity. The analogy between Friedmann and Vialov equations suggests that the Friedmann equation applies only when $a(t) \ll t$, or when the scale factor $a(t)$ is much smaller than the age of the universe t at that time, which means away from the Big Bang. This naive suggestion agrees with the idea of the existence of a fundamental length, the Planck length ℓ_{pl} below which the description of spacetime with a continuous manifold breaks down.

The property that all solutions of the Friedmann equation are roulettes transfers to the glaciology side of the analogy and to the solutions of the rescaled Vialov equation, which are seen to all be roulettes. Also the symmetry of the spatially flat Einstein–Friedmann equations [151]

$$a \to \tilde{a} = a^s\,, \tag{6.118}$$

$$dt \to d\tilde{t} = s\, a^{\frac{3(w+1)(s-1)}{2}} dt\,, \tag{6.119}$$

$$\rho \to \tilde{\rho} = a^{-3(w+1)(s-1)} \rho\,, \tag{6.120}$$

with $s \neq 0$ is mirrored by a symmetry of the rescaled Vialov equation (6.108), expressed by

$$h \to \tilde{h} = h^s\,, \tag{6.121}$$

$$d\bar{x} \to d\tilde{x} = s\, h^{\frac{2(n+1)(s-1)}{n}} d\bar{x}\,, \tag{6.122}$$

as can be checked directly.

Let us discuss now the limit of perfectly plastic ice $n \to +\infty$. Using $\lim_{n \to +\infty} n^{1/n} = 1$, the original Vialov equation (in terms of x) becomes

$$h \, |h'| \, h^{2/n} = \frac{(n+2)^{1/n} \left[x \, c(x) \right]^{1/n}}{(2\mathcal{A})^{1/n} \, \rho_{\text{ice}} \, g} \to \frac{1}{\rho_{\text{ice}} \, g} . \qquad (6.123)$$

The differential equation describing the longitudinal glacier profile assuming perfectly plastic ice is

$$h \, |h'| = \frac{\tau_b}{\rho_{\text{ice}} \, g} , \qquad (6.124)$$

where the constant τ_b is the stress at the bottom of the ice, and is derived by using Eq. (6.106) with the Lagrangian $\mathcal{L}_\infty$. The limit $n \to +\infty$ of the symmetry produces

$$h \to \tilde{h} = h^s , \qquad (6.125)$$

$$dx \to d\tilde{x} = s \, h^{2(s-1)} \, dx \qquad (6.126)$$

(it leaves invariant the left hand side of Eq. (6.124), while the right hand side is invariant because it is constant). The solution of Eq. (6.124) is the well–known parabolic Nye profile [302]

$$h(x) = H \sqrt{1 - \frac{x}{L}} , \qquad (6.127)$$

$$H = \sqrt{\frac{2 \, \tau_b \, L}{\rho_{\text{ice}} \, g}} . \qquad (6.128)$$

This parabola is also a roulette: in geometry the parabola is obtained as a point fixed on the involute of the circle as the latter rolls along a straight line.

6.4.4 *Nye profile as a generating function*

The essence of the symmetry of the Friedmann equation for spatially flat FLRW universes is that the solution corresponding to a perfect fluid can be obtained from that for another fluid. This is true also on the glaciology side of the analogy, where the solution of the Vialov equation for general n can be generated starting from a known solution for a particular value of this parameter. The simplest solution is the Nye profile for perfectly plastic ice ($n \to +\infty$) and it is natural to reduce the problem of solving the Vialov equation for finite n to this case [149]. The Vialov equation can, indeed, be reduced to the Nye equation (6.124) through one more variable change [149].

We retain $\bar{x}$ (defined in Eq. (6.98)) as the independent variable, but we change the dependent variable:

$$h \to \bar{h} = h^p , \qquad (6.129)$$

which transforms the Vialov equation (6.100) into

$$\bar{h}^{-\frac{2(n+1)-np}{np}} \frac{d\bar{h}}{d\bar{x}} = \frac{p}{\alpha^{1/n}} . \qquad (6.130)$$

The specific choice

$$p = \frac{n+1}{n} \qquad (6.131)$$

of the exponent in Eq. (6.129) then changes the Vialov equation into

$$\bar{h} \left| \frac{d\bar{h}}{d\bar{x}} \right| = \frac{1}{\bar{\alpha}^{1/n}} \qquad (6.132)$$

for $\bar{h}(\bar{x})$, where $\bar{\alpha} = \left(\dfrac{n}{n+1}\right)^n \alpha$. This is the Nye equation. This transformation can, at least in principle, be used to generate solutions of the Vialov equation.

Let us examine again, as an example, the Vialov model with given accumulation function $c(x) = c_0 = \text{const.}$, for which we have derived the rescaled variable

$$\bar{x} = \frac{n \, c_0^{1/n}}{n+1} x^{\frac{n+1}{n}} ; \qquad (6.133)$$

introducing $\bar{h} = h^{\frac{n+1}{n}}$, the Vialov equation reduces to the Nye equation which has the "barred" Nye profile

$$\bar{h}(\bar{x}) = \bar{H} \sqrt{1 - \frac{\bar{x}}{\bar{L}}} \qquad (6.134)$$

as the solution. By simply rewriting the latter in terms of the "unbarred" quantities h and x, we have

$$h^{\frac{n+1}{n}} = \bar{H} \sqrt{1 - \frac{n \, c_0^{1/n}}{(n+1)\bar{L}} x^{\frac{n+1}{n}}} . \qquad (6.135)$$

We then raise both sides to the power $\dfrac{n}{n+1}$ and use the constants

$$H = \bar{H}^{\frac{n+1}{n}}, \tag{6.136}$$

$$L = \left[\frac{(n+1)\,\bar{L}}{n\,c_0^{1/n}} \right]^{\frac{n}{n+1}}, \tag{6.137}$$

to rewrite the Nye solution (6.134). The result is the Vialov solution

$$h(x) = H\left[1 - \left(\frac{x}{L}\right)^{\frac{n+1}{n}} \right]^{\frac{n}{2(n+1)}}. \tag{6.138}$$

There is an important *caveat*, however: in the general case in which the accumulation function $c(x)$ is not constant, one may not be able to express explicitly $\bar{x}$ in terms of elementary functions of x. If such an expression is impossible, the reduction of the Vialov equation to the Nye equation is purely formal and generating its solution from the generating function (6.134) is only a formal theoretical possibility. The forms of the accumulation function $c(x)$ for which the integral $I(x)$ is computed explicitly in terms of elementary functions are identified in Ref. [140] using the Chebyshev theorem. To conclude, the cosmological analogy unveils a surprising similarity between the shape of glaciers and ice caps and spatially flat FLRW universes and an elegant, albeit formal, reduction of Glen law ice and Vialov shapes to perfectly plastic ice and Nye parabolas.

The reader will have noted that there is also some similarity between the equation for equilibrium beach profiles and the Vialov equation: the rescaled Vialov equation is

$$\left(\frac{dh}{d\bar{x}}\right)^2 = \frac{h^{\frac{-2(n+2)}{n}}}{\alpha^{2/n}}, \tag{6.139}$$

while the equation for equilibrium beach profiles instead is inhomogeneous:

$$\left(\frac{dh}{dx}\right)^2 = C^2\,h^{\frac{-3(m+1)}{2}} - 1. \tag{6.140}$$

As we now know from FLRW cosmology, the physical and mathematical differences introduced by the inhomogeneous curvature term are significant.

Chapter 7

Earthquakes and their cosmic cousins

Analogies are not "aids" to the
establishment of theories; they
are an utterly essential part of
theories, without which theories
would be completely valueless
and unworthy of the name.

Norman Robert Campbell

Earthquakes are among the most devastating phenomena occurring on Earth: they can change the landscape by destroying landmarks, changing the course of rivers, or creating rifts, not to mention destroying buildings, bridges, and roads. Earthquakes are responsible for most of the destruction of stone monuments and buildings from antiquity (Fig. 7.1).

It is rather surprising that one of the best–known results of seismology, the Omori law, lends itself to an analogy with FLRW cosmology in which the main shock of an earthquake has a counterpart in the Big Rip of the universe.

7.1 Omori's law

The Omori law says that, on average, the number of aftershocks per unit time $n(t)$ following an earthquake's main shock decays according to the empirical power law [305]

$$n(t) = \frac{k}{c+t} = \frac{k}{t - |c|},$$ (7.1)

where $k > 0$ and $c < 0$ are constants and $t > |c|$. The Omori law describes also the seismicity rate before and after a volcanic eruption[1] [256, 365].

In spite of a large literature on Omori's law (see the reviews [192, 412]), its physical origin and interpretation remain mysterious. The cause of the earthquake shocks should be due to a rupture mechanism in the rocks composing the Earth's crust, but the mechanism is unclear. Although the main approach to the study of earthquakes is phenomenological, there is reason to believe the Omori law to be more fundamental than a mere data–fitting device and there is hope to derive it from basic models. One such attempt, proposed in Refs. [190–192] starts out deriving the first order differential equation satisfied by $n(t)$,

$$\dot{n}(t) = -\sigma \, n^2(t), \tag{7.2}$$

where $\sigma = k^{-1}$ and an overdot denotes differentiation with respect to time. Refs. [190–192] make an analogy between the decaying number of

---

[1]To describe the beforeshock phase, one needs to introduce an absolute value, $n(t) = k\left|t - |c|\right|^{-1}$.

FIG. 7.1. The Doric temple of Hera (temple E) in Selinunte, Sicily. Most of the ancient Greek buildings in this seismic zone have been damaged or destroyed by earthquakes over the centuries.

aftershocks per unit time and the phenomenology of decreasing density of ions in the ionospheric plasma due to the recombination of opposite charges [190–192]. Let $n_\pm$ be the density of positive or negative charges in this plasma, then the total number of charges is

$$n(t) = n_+ + n_-\tag{7.3}$$

and the recombination equation is

$$\dot{n} = -\sigma\, n_+ n_- \, .\tag{7.4}$$

The key observation is that, in a globally neutral plasma, $n_+ \simeq n_-$ and the recombination equation is approximately the same as Eq. (7.2). The existence of positive/negative pairs suggests an analogy in which an earthquake occurs due the fast slipping of rock along a fault plane in the Earth's crust. There are two adjacent sides in a tectonic fault, their numbers being denoted by n_+ and n_-, respectively. When an active fault ruptures, energy is released and the stresses on its parallel sides are neutralized, reducing the number n of active faults, similar to a pair of oppositely charged ions that neutralize each other in the ionospheric plasma. Lending credibility to this analogy, the evolution of the number of faults should follow the law [190–192]

$$\frac{dn}{dt} = -\sigma\, n_+ n_- \simeq -\sigma\, n^2\tag{7.5}$$

where σ is a constant deactivation coefficient and $n_+ = n_-$. In this suggestive analogy, the fact that adjacent fault sides are neutralized in pairs excludes different powers in the Omori law [191, 192]. In other descriptions of aftershocks, the deactivation coefficient σ could depend on time, which introduces non–stationarity and deviations from the Omori law [191, 192].

In addition to the aftershock phase, often also the beforeshock phase, during which secondary shocks increase their frequency preceding the main shock, is described by a version of Omori's law

$$\dot{n} = \sigma\, n^2 \, ,\tag{7.6}$$

although the phenomenological descriptions and the data–fitting are different.

The differential equation satisfied by Omori's law is reminiscent of the properties of the Friedmann equation for FLRW universes dominated by a phantom fluid causing a Big Rip at a finite time. The Big Rip separates "before" and "after" universes and is analogous to the main earthquake shock. This analogy is intriguing and may provide some insight about variability of the deactivation coefficient σ versus variability of the power in the Omori law. As usual, the cosmological analogy offers also a solution of the inverse variational problem for Eq. (7.5) and for its generalizations.

7.2   Mechanical analogy and Lagrangian formulation

Predictably, the seismological literature does not discuss Lagrangians in the description of earthquakes, but it is not difficult to guess a Lagrangian leading to Omori's law [146]

$$L\left(n, \dot{n}\right) = n\,\dot{n}^2 + \sigma^2 n^5\,. \tag{7.7}$$

The corresponding Euler–Lagrange equation

$$\frac{d}{dt}\left(\frac{\partial L}{\partial \dot{n}}\right) - \frac{\partial L}{\partial n} = 0 \tag{7.8}$$

reads

$$2\,n\,\ddot{n} + \dot{n}^2 - 5\sigma^2 n^4 = 0\,. \tag{7.9}$$

The Omori law (7.2) is a first integral of Eq (7.9), as is evident from the fact that the differentiation of (7.2) yields

$$\ddot{n} = -2\,\sigma\,n\,\dot{n} = 2\,\sigma^2\,n^3\,. \tag{7.10}$$

In conjunction with (7.2), this leads to $2\,n\,\ddot{n} + \dot{n}^2 - 5\sigma^2 n^4 = 0$.

The Hamiltonian corresponding to the Lagrangian (7.7) is

$$\mathcal{H} = \pi_n\,\dot{n} - L = n\left(\dot{n}^2 - \sigma^2 n^4\right)\,, \tag{7.11}$$

where $\pi_n = \partial L/\partial \dot{n} = 2\,n\,\dot{n}$ is the momentum conjugated to n. Since this Hamiltonian does not depend explicitly on the time t, it is conserved. By substituting the Omori law (7.2) in the Hamiltonian (7.11), one obtains

$$\mathcal{H} = 0 : \tag{7.12}$$

the point–particle analogous to the Omori system has conserved zero mechanical energy. The energy integral can be written as

$$\frac{\mathcal{H}}{2} = \mu\left(\frac{\dot{n}^2}{2} - \frac{\sigma^2}{2} n^4\right) \tag{7.13}$$

where, for $n \geq 0$, $\mu(n) = n$ can be viewed as a position–dependent mass for an effective analogous particle. The corresponding kinetic energy is $\mu\,\dot{n}^2/2$ and the particle moves in the potential energy

$$V(n) = -\frac{\mu\,\sigma^2 n^4}{2} \tag{7.14}$$

with zero total mechanical energy. The aftershock phase of an earthquake has $\dot{n} < 0$ and its analogous particle moves to the left of the n–axis toward $n = 0$, reflecting the fact that aftershocks are more frequent near the initial point $n_{(0)} > 0$ and stop as $n \to 0$.

A very simple phase plane $\left(n, \dot{n}\right)$ is associated with Omori's law. The energy constraint (7.2) (or, equivalently, Eq. (7.12)) forces the representative points of the system to move on the parabolas $\dot{n} = \mp \sigma\, n^2$. The upper sign describes the aftershock phase, while the lower one corresponds to the beforeshock phase. These two parabolas are the only orbits of the two different dynamical systems, which are pictured as living in the same phase plane only for convenience. Although they seem to touch each other at the origin $(0,0)$, in reality they are disconnected curves.

The aftershock phase lives in the lower quadrant $n \geq 0\,, \dot{n} \leq 0$ in which the point representing the state of the system in this phase plane is attracted by the origin and moves along the downward–facing parabola towards it, regardless of where it starts from in this quadrant. In this regime, the secondary shocks decay in a finite time $|c|$.

The beforeshock phase, instead, corresponds to the upper quadrant $n \geq 0\,, \dot{n} \geq 0$. Here, the point representing the dynamical system moves away from the origin and upward toward infinite n and $\dot{n}$, reaching infinity in a finite time. The main shock corresponds to the pole $t = |c|$ in the solution

$$n(t) = \frac{k}{|t - |c||} \,. \tag{7.15}$$

The dynamics is discontinuous at the main shock: the solutions describing beforeshock and aftershock are disconnected.

7.3 Analogy with the Big Rip

Let us discuss now the analogy with FLRW cosmology. We square Eq. (7.2) and rewrite it as

$$\left(\frac{\dot{n}}{n}\right)^2 = \sigma^2 n^2 \,, \tag{7.16}$$

then the analogy with the Friedmann equation (1.179) becomes obvious if the scale factor $a(t)$ corresponds to the function $n(t)$. The analogous FLRW universe is spatially flat and is permeated by a fluid with energy density

$$\rho(t) = \rho_0 \, a^2(t) \,, \tag{7.17}$$

as implied by Eq. (7.16), where ρ_0 is a positive integration constant and

$$\sigma^2 = \frac{8\pi G \rho_0}{3} \,. \tag{7.18}$$

The analogy is complete if

$$\rho(a) = \frac{\rho_0}{a^{3(w+1)}} \tag{7.19}$$

with $w = -5/3$. The aftershock regime is analogous to a contracting universe with decreasing $a(t)$ and $\dot{n} < 0$, while the beforeshock phase corresponds to an expanding universe ($\dot{n} > 0$). In the aftershock phase ($\dot{n} < 0$), the analogous Friedmann equation describes a spatially flat contracting universe containing a phantom fluid with energy density $\rho = \rho_0\, a^2$ and equation of state $P = -5\rho/3$. A phantom fluid causes the universe to expand so fast that it explodes at a finite time in a Big Rip [72,73]. In a Big Rip the scale factor $a(t)$ does not go to zero but diverges, together with the scalar curvature invariants and the energy density ρ and pressure P [72,73].

In the analogy, $t = |C|$ corresponds to the main earthquake shock and is analogous to the Big Rip, while the aftershock phase $\dot{n} < 0$ corresponds to the branch of a universe contracting from a Big Rip. The expanding and contracting branches on either side of the Big Rip are disconnected because the spacetime manifold ends at a curvature singularity [435]. The expanding branch of the phantom–dominated universe is analogous to the Omori law with sign changed $\dot{n} = \sigma\, n^2$ modelling the beforeshock phase of an earthquake, during which smaller shocks become more and more frequent and lead to the main shock [256, 365]. The main earthquake shock corresponds to the Big Rip.

The mathematical reason for the catastrophic nature of the solution of Eq. (7.2) and of $\dot{a} \approx a^2$ is the exponent 2 in the right hand side: it prevents the existence of a maximal solution defined on an infinite half–interval [61]. The analogy may have heuristic value for physical insight. The Lagrangian (7.7) reproducing the Omori law is inspired by the effective point–like Lagrangian for the FLRW universe [75, 76, 106, 133, 358]

$$L\,(a, \dot{a}, P) = 3\,a\,\dot{a}^2 - a^3 P\,. \tag{7.20}$$

Another Lagrangian for Omori's law is [177]

$$\mathcal{L}\,(t, \dot{n}) = \frac{1}{2}\left(t + c\right)^2 \dot{n}^2\,. \tag{7.21}$$

Curiously, the Hamiltonian coincides with the Lagrangian,

$$\mathcal{H} = \dot{n}\,\frac{\partial \mathcal{L}}{\partial \dot{n}} - \mathcal{L} = \mathcal{L} \tag{7.22}$$

and is not conserved. Since $\mathcal{L}$ does not depend on n, the canonically conjugated momentum $\pi_n = (t + c)^2\,\dot{n}$ is constant, which reproduces Eq. (7.2) (but not Eq. (7.6)). However, because it depends explicitly on time, this Lagrangian does not lend itself to a cosmological analogy.

7.4   Generalizations

Generalizations of the Omori law exist in the literature, in which aftershocks are often modelled by the phenomenological Omori–Utsu law [211, 234]

$$n(t) = \frac{k}{\left(t - |c|\right)^p} , \qquad (7.23)$$

where $t > |c|$ and the exponent p varies in a rather wide range depending on the location and even on the specific earthquake event [411]. Again, one writes the analogue of Eq. (7.2) as

$$\dot{n} = -\frac{p}{k^p} n^{\frac{p+1}{p}} \qquad (7.24)$$

and derives the Lagrangian

$$L_{(p)}\left(n, \dot{n}\right) = n \dot{n}^2 + \sigma_{(p)}^2 \, n^{\frac{3p+2}{p}} . \qquad (7.25)$$

The corresponding Euler–Lagrange equation is

$$2 \, n \ddot{n} + \dot{n}^2 - \frac{(3p+2)}{p} \sigma_{(p)}^2 \, n^{\frac{2(p+1)}{p}} = 0 \qquad (7.26)$$

and the Hamiltonian reads

$$\mathcal{H}_{(p)} = n \left(\dot{n}^2 - \sigma_{(p)}^2 \, n^{\frac{2(p+1)}{p}} \right) . \qquad (7.27)$$

Again, the Hamiltonian is conserved and $\mathcal{H}_{(p)} = 0$ reproduces the equation of motion. There is still a cosmological analogy. For the range of values of $p > 0$ found in the literature, the FLRW fluid remains a phantom fluid with equation of state $P = w_p \rho$ and

$$w_p = -\frac{(3p+2)}{3p} . \qquad (7.28)$$

The Big Rip, which occurs for all equation of state parameters $w < -1$, is still present. The fact that the exponent p deviates from unity spoils the analogy with a plasma of Refs. [190–192]. These authors attribute the deviation from the $p = 1$ Omori law to the time dependence of the coefficient σ. In the cosmological analogy, a time–dependent $\sigma(t)$ corresponds to a time–varying gravitational constant G (see Eq. (7.18)), which is impossible in Einstein gravity. However, a varying gravitational coupling occurs in scalar–tensor gravity, which implies modifying the Einstein–Friedmann equations [60, 77, 133, 174]. If a cosmological analogy still applies, it would imply that the time variation of σ brings in new energy terms dependent on $\dot{\sigma} \neq 0$ in the energy balance involving the variation of n. In cosmology, a less drastic alternative consists of modifying the equation of state of the cosmic fluid, perhaps replacing it with a dynamical equation of state. This phenomenology can be achieved by replacing the cosmic fluid with a scalar field minimally coupled with the curvature, which still behaves as a perfect fluid [5,193,244,260,264,384]. Correspondingly, the fundamental derivation of the Omori law in Refs. [190–192] would change. Extra ingredients could involve a distribution of intersecting faults with more than two adjacent sides involved.

Chapter 8

Miscellaneous cosmic analogies

> *If genius has any common
> denominator, I would propose
> breadth of interest and the
> ability to construct fruitful
> analogies between fields.*

Stephen Jay Gould

8.1 Introduction

There are many more analogies with cosmology in physics and in the natural sciences, in addition to those already discussed in some detail in the previous chapters. Some of these analogies may be of interest to some readers and marginal to others, some may be mere curiosities and others could disclose features of physical systems that were previously overlooked. In this last chapter, we sketch a few more of these analogies without entering the discussion in any great detail. We report also some analogies with fluids of negative density, which are not of physical interest for cosmology but may lead to formal properties that are useful on the other side of the analogy. The analogous systems and phenomena discussed in this chapter include the build–up of electric charge on ice pellets in a thunderstorm cloud, capillary flows, systems ruled by surface tension, the dispersion relation for electromagnetic waves in a plasma, the fronts of lava flows in volcanic eruptions, and river floods. Many more analogies no doubt exist and we encourage the reader to search for them, which should not be too difficult given the familiarity with the Einstein–Friedmann equations (1.179), (1.180), and (1.187) acquired by now.

8.2   Charge buildup in a thunderstorm cloud

The physics of thunderstorms is complicated, but it can be summarized by saying that charges in a thundercloud are generated and separated when supercooled droplets graze the undersides of hail pellets, which are polarized by fair weather electric fields. Droplets rebound and carry away a positive charge Q, leaving a negative charge $-Q$ on the hail pellet. The positively charged droplets are then carried upwards in the cloud by updraught while the negatively charged hail pellets fall towards the base of the cloud. The latter becomes a giant dipole and can discharge part of its charge through lightning. The thunderstorm cell recharges and can be seen as a giant generator. The story is much more complex, of course: see Refs. [13, 178, 275, 276] for the real story and Ref. [277] for a simplified account.

Let $Q(t)$ be the charge building up on an ice pellet, R be the radius of the pellet, ϑ the angle between the trajectory of a grazing droplet and the radial direction to the centre of the pellet (assumed to be spherical) at the tangent point. Let F be the electric field, directed downward in the thunderstorm cloud, and τ be a time scale. Then, the time rate of charging of a hail pellet is [277]

$$\frac{dQ}{dt} = -\frac{1}{\tau}\left(3FR^2\cos\vartheta + Q\right) . \tag{8.1}$$

We can write this equation simply as

$$\frac{\dot{Q}}{Q} = -\left(\frac{\alpha}{Q} + \beta\right) , \tag{8.2}$$

with the constants

$$\alpha = \frac{3FR^2}{\tau}\cos\vartheta , \tag{8.3}$$

$$\beta = \frac{1}{\tau} . \tag{8.4}$$

Squaring both sides gives

$$\left(\frac{\dot{Q}}{Q}\right)^2 = \beta^2 + \frac{\alpha^2}{Q^2} + \frac{2\alpha\beta}{Q} , \tag{8.5}$$

which is analogous to the Friedmann equation

$$H^2 = \frac{\Lambda}{3} - \frac{K}{a^2} + \frac{8\pi G\rho_0}{3a} \tag{8.6}$$

with a positive cosmological constant and negative curvature index K. The covariant conservation equation (1.187) is automatically satisfied by the fluid with energy density $\rho(a) = \rho_0/a$ which has equation of state parameter $w = -2/3$ corresponding to a universe dominated by frustrated domain walls [34, 421].

The solution of Eq. (8.2) is the exponential [277]

$$Q(t) = -3FR^2 \cos\vartheta \left(1 - e^{-t/\tau}\right) \tag{8.7}$$

and the analogous open universe is contracting.

The Lagrangian of the analogous FLRW universe provides the Lagrangian for the charging phenomenon

$$L\left(Q, \dot{Q}\right) = 3Q\dot{Q}^2 + 6\alpha\beta\, Q^2 + 3\beta^2 Q^3 + 3\alpha^2 Q\,. \tag{8.8}$$

Since L does not depend explicitly on time, the corresponding Hamiltonian

$$\mathcal{H} = \dot{Q}\frac{\partial L}{\partial \dot{Q}} - L = 3Q\dot{Q}^2 - 6\alpha\beta\, Q^2 - 3\beta^2 Q^3 - 3\alpha^2 Q \tag{8.9}$$

is constant. The original equation (8.2) is recovered by setting this constant to zero and taking the negative sign in the resulting square root, which corresponds to a contracting analogous universe and to the droplets carrying away positive charge from the hail pellet. The FLRW symmetries do not apply to this negatively curved analogous universe and to Eq. (8.2).

8.3 de Sitter space in a meniscus

Consider a vertical fiber of uniform radius b that is immersed in a liquid with surface tension γ and is wetted by this liquid [110]. When equilibrium between the liquid bath and the liquid in the fiber is reached,[1] the vertical components of the tension force (due to the liquid bath and to the liquid in the fiber) must balance. Let x be a vertical axis along the fiber and pointing downward and let $z(x)$ describe the free surface of the liquid climbing this fiber. Then the angle ϑ between the profile $z(x)$ of this free surface and the vertical is given by $\tan\vartheta = dz/dx \equiv z'(x)$. The balance of forces in the vertical direction reads

$$2\pi\, z\gamma\cos\vartheta = 2\pi\gamma\, b\,. \tag{8.10}$$

Using

$$\cos\vartheta = \frac{1}{\sqrt{1 + \tan^2\vartheta}} = \frac{1}{\sqrt{1 + (z')^2}}\,, \tag{8.11}$$

---

[1] The dynamical approach to equilibrium is studied in Ref. [90] and is of interest in designing devices to measure surface tension or in welding and soldering processes using surface coating methods.

the force balance (8.10) gives [110]

$$\frac{z}{\sqrt{1 + (z')^2}} = b \,, \tag{8.12}$$

which is recognized as the differential equation (2.169) encountered in the catenary problem and is rearranged as the Friedmann–like equation

$$\left(\frac{z'}{z}\right)^2 = \frac{1}{b^2} - \frac{1}{z^2} \,. \tag{8.13}$$

The cosmological analogy $z(x) \rightarrow a(t)$ involves the positive cosmological constant $\Lambda = 3/b^2$ and the positive curvature index $K = 1$; the analogous universe is empty and has positive spatial curvature and positive cosmological constant. The solution of Eq. (8.13) is the catenary $z(x) = b \cosh(x/b)$ corresponding to the scale factor

$$a(t) = \sqrt{\frac{3}{\Lambda}} \cosh\left(\sqrt{\frac{\Lambda}{3}} t\right) \,. \tag{8.14}$$

A rising meniscus formed by a liquid wetting the wall of a container is approximately described by the same differential equation and assumes the same catenary form.

8.4 Other fluids ruled by surface tension

There are potentially other analogies between FLRW cosmology and fluid systems ruled by surface tension, without including the well known fluid–dynamical systems of analogue gravity used to investigate Hawking radiation, superradiance, and other phenomena of quantum field theory in curved spacetime that we have mentioned briefly at the beginning of this book [58, 324, 325, 368, 405, 407, 408, 425, 442].

8.4.1 *Dynamical bubbles*

A connection between the Rayleigh–Plesset equation describing bubbles with time–dependent radius in a viscous fluid and the Friedmann equation was established in Refs. [271, 348]. Another partial analogy between cosmology and the dynamics of a bubble with time–dependent radius is proposed in Ref. [350]. In an incompressible viscous fluid, consider a bubble that can expand or contract and let $R(t)$ be its radius at time t. Let ρ_ℓ, ν, and γ be the density, kinematic viscosity coefficient, and surface tension of this fluid, respectively, while $\Delta P(t) = P_{\text{in}}(t) - P_{\text{out}}(t)$ is the pressure difference between the interior and the exterior of the bubble. Then the dynamics of the bubble is ruled by the extended Rayleigh–Plesset equation [62, 171, 332, 340]

$$\frac{\ddot{R}}{R} + \frac{3}{2}\left(\frac{\dot{R}}{R}\right)^2 - \frac{\Delta P(t)}{\rho_\ell R^2} + \frac{2\gamma}{\rho_\ell R^3} + 4\nu\,\frac{\dot{R}}{R^3} = 0\,. \tag{8.15}$$

A similar equation can be derived in the cosmology of a universe filled with a dust of density $\rho = \rho_0/a^3$ and with $\Lambda = 0$ and negative curvature index K. Multiplying the Friedmann equation (1.179) by $3/2$ and adding it to the acceleration equation (1.180), one obtains

$$\frac{\ddot{a}}{a} + \frac{3}{2}\frac{\dot{a}^2}{a^2} + \frac{3K}{2\,a^2} - \frac{8\pi G\rho_0}{3\,a^3} = 0\,. \tag{8.16}$$

The first two terms in the extended Rayleigh–Plesset equation (8.15) have FLRW analogues in Eq. (8.16), where the bubble radius $R(t)$ obviously corresponds to the scale factor $a(t)$. However, in order to have a full analogy, one needs a negative dust density with $\rho_0 = -\dfrac{3\gamma}{4\pi G\rho_\ell}$ and the curvature index $K = -\dfrac{2\,\Delta P(t)}{3\rho_\ell}$ should be approximated with its constant average, while the viscosity ν must be negligible. While the analogy is incomplete, bubbles are easy systems to realize in the laboratory and one can in principle play with various physical quantities to vary the conditions of the fluid in order to pursue cosmic analogies [350].

8.4.2 *Capillary flow*

A potential analogy between FLRW cosmology and the capillary flow of an inviscid fluid (possibly a superfluid) has been suggested in Ref. [41]. Consider a fluid of density ρ_ℓ, surface tension γ, and kinematic viscosity ν in a capillary tube of diameter h, and let θ be the contact angle. If the fluid wets the walls of the pipe (hydrophilic behaviour), it is $0 < \theta < \pi/2$ while, for hydrophobic behaviour, $\pi/2 < \theta < \pi$. Let z be an axis along the capillary tube. The moving front of the fluid has centerline position $z(t)$, where t is the time. Newton's second law of motion applied to a moving element of the fluid yields the Washburn equation [437]

$$\frac{d}{dt}\left(z\,\frac{dz}{dt}\right) = -\eta\,z\,\frac{dz}{dt} + \frac{2\gamma}{\rho_\ell\,h}\,\cos\theta\,, \tag{8.17}$$

where $\eta = 12\nu/h^2$ is the inverse of a time scale and, in general, the surface tension $\gamma = \gamma_0\, \chi(z)$ (where γ_0 is a constant) and the contact angle θ depend on the position z along the tube. The cosmological analogy arises by adding the Friedmann equation (1.179) and the acceleration equation (1.180) for a perfect fluid with equation of state $P = w\,\rho$ and density $\rho = \rho_0/a^{3(w+1)}$, and multiplying the resulting equation by a^2. One obtains

$$\dot{a}^2 + a\,\ddot{a} = \frac{d}{dt}\left(a\,\frac{da}{dt}\right) = \frac{4\pi G}{3}\,(1 - 3w)\,\rho_0\, a^{-(3w+1)} - K + \frac{2\Lambda}{3}\,a^2\,. \quad (8.18)$$

By tuning the form of the functions $\chi(z)$ and $\cos\left(\theta(z)\right)$ in the laboratory, one could potentially make the Washburn equation (8.17) for an inviscid fluid with $\nu = 0$ (possibly a superfluid) mimic Eq. (8.18) and then the capillary flow mimics FLRW universes [41].

8.5   Analogies with negative density fluids

Several physical systems are ruled by equations that provide analogies with the Friedmann equation (1.179), but the matter source contains at least one fluid with negative energy density, which is unphysical in cosmology. The formal analogy still stands and is useful if one is interested in solving the inverse variational problem and finding Lagrangians and Hamiltonians, or in finding symmetries of the system (assuming that the analogous universe is spatially flat). We report below a few of these systems.

8.5.1   *Plasma dispersion relation*

Electromagnetic waves with angular frequency ω and wave vector $\vec{k}$ propagating in a plasma obey the dispersion relation

$$\omega(k) = \sqrt{c^2 k^2 + \omega_p^2}\,, \quad (8.19)$$

where we have restored the speed of light c and the constant ω_p is the plasma frequency. The phase velocity of these waves is

$$v_p = \frac{\omega}{k} = \sqrt{c^2 + \frac{\omega_p^2}{k^2}}\,, \quad (8.20)$$

while their group velocity is

$$v_g = \frac{d\omega}{dk} = \frac{c}{\sqrt{1 + \frac{\omega_p^2}{c^2 k^2}}} = \frac{c^2 k}{\sqrt{c^2 k^2 + \omega_p^2}}\,. \quad (8.21)$$

We can write

$$\left(\frac{1}{\omega}\frac{d\omega}{dk}\right)^2 = \left(\frac{v_g}{\omega}\right)^2 = \frac{c^4 k^2}{\left(c^2 k^2 + \omega_p^2\right)\omega^2} \tag{8.22}$$

and, eliminating k with Eq. (8.19),

$$\left(\frac{1}{\omega}\frac{d\omega}{dk}\right)^2 = \frac{c^2}{\omega^2} - \frac{c^2\,\omega_p^2}{\omega^4}. \tag{8.23}$$

If we identify k with the comoving time t and $\omega(k)$ with the scale factor $a(t)$, this equation is formally analogous to the Friedmann equation

$$H^2 = -\frac{K}{a^2} + \frac{8\pi G\rho_0}{3\,a^4} \tag{8.24}$$

which describes a universe with negative curvature index $K = -c^2$, zero cosmological constant, and a radiation fluid with (unphysical) negative energy density since $8\pi\rho_0/3 = -c^2\omega_p^2$. The corresponding scale factor $a(t) = a_0\sqrt{t^2 + t_0^2}$ describes a bouncing universe contracting from infinite size to a minimum size $a_{\min} = a_0 t_0$, where it bounces and then expands forever. A singularity is avoided thanks to the violation of the energy conditions by the negative density radiation fluid.

The effective Lagrangian and Hamiltonian for the dispersion relation (8.19) inspired by FLRW cosmology read (using the notation $\omega' \equiv d\omega/dk$)

$$L\left(\omega,\omega'\right) = \omega(\omega')^2 - \frac{c^2\,\omega_p^2}{\omega} + c^2\omega\,, \tag{8.25}$$

$$\mathcal{H} = \omega\,(\omega')^2 + \frac{c^2\,\omega_p^2}{\omega} - c^2\omega\,, \tag{8.26}$$

and setting $\mathcal{H} = 0$ reproduces Eq. (8.22). The equation $\mathcal{H} = 0$, corresponding to the conservation of the Hamiltonian and to the vanishing of this constant "energy" of the problem has a physical meaning. It can be written as

$$v_g^2 + \frac{c^2}{\omega^2}\left(\omega^2 - c^2 k^2\right) = c^2\,, \tag{8.27}$$

where we have used the dispersion relation (8.19) to eliminate ω_p. Then, recognizing k/ω as $1/v_p$, we can write $v_g^2\,v_p^2 = c^4$ and

$$\sqrt{v_p\,v_g} = c\,, \tag{8.28}$$

the well–known result that the geometric mean of the phase and group velocities in a plasma is the speed of light *in vacuo* (Fig. 8.1).

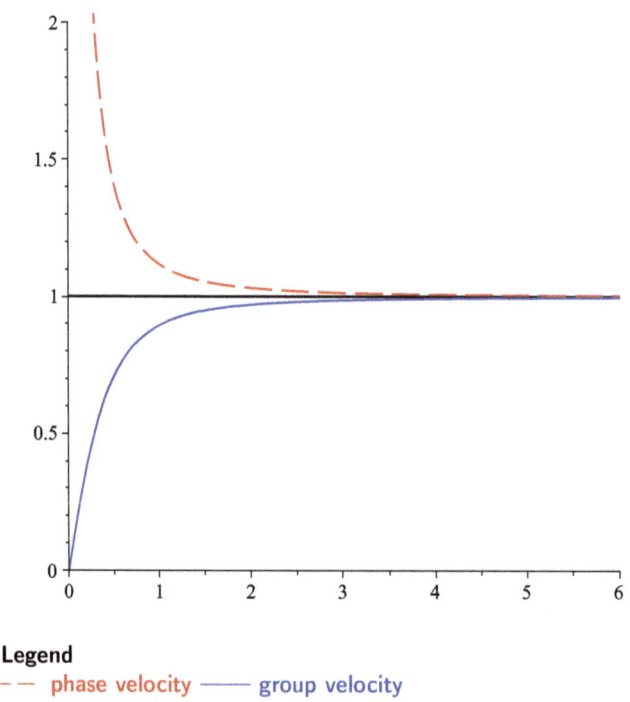

Legend
-- phase velocity ——— group velocity
——— c, the geometric mean of v_p and v_g

FIG. 8.1. The phase velocity v_p and the group velocity v_g as functions of the wave vector k (in units of the speed of light c *in vacuo*).

8.5.2   *Fronts of lava flows*

As the front of a lava flow advances, it buries whatever is on its path and sets on fire the surroundings, before finally cooling down, beginning from a crust that thickens until all the lava is solidified (Fig. 8.2).

An advancing lava flow has a front that can be modelled in a way similar to a longitudinal glacier profile, although the material is different (see Ref. [187] for a review). Assuming that the lava front moves at constant velocity on a sloping plane, that it has infinite extent in the transversal direction, and that it is homogeneous and isothermal, one denotes by x the longitudinal coordinate in the direction of the flow and by $h(x)$ the lava thickness, measured along an axis perpendicular to the bed (this quantity is similar to the ice thickness describing longitudinal glacier profiles). If the length of the flow is L, assume $h \ll L$ (which parallels the shallow ice ap-

FIG. 8.2. Lava flow at the bottom of Mt. Etna, Italy

proximation in glaciology). Let ρ be the density of the lava, η its dynamic viscosity coefficient, g be the acceleration of gravity, ϑ be the angle of the lava bed with the horizontal, and v_0 the velocity of the lava at the bed. Then, treating the lava as a Newtonian fluid,[2] the Navier–Stokes equations for laminar flow give the non-linear ordinary differential equation for the lava front profile $h(x)$

$$\beta\, h^3 h' - \alpha\, h^3 + v_0\, h = 0 \qquad (8.29)$$

where $h' \equiv dh/dx$ and

$$\alpha = \frac{\rho\, g \sin\vartheta}{3\eta}, \qquad (8.30)$$

$$\beta = \frac{\rho\, g \cos\vartheta}{3\eta}. \qquad (8.31)$$

We can easily form an analogy with FLRW cosmology: isolating h' and rewriting Eq. (8.29) as

$$\left(\frac{h'}{h}\right)^2 = \frac{\alpha^2}{\beta^2\, h^2} + \frac{v_0^2}{\beta^2\, h^6} - \frac{2\alpha\, v_0}{\beta^2 h^4}\,; \qquad (8.32)$$

_____

[2]There is, however, evidence that lava is better described as a non–Newtonian Bingham fluid [115].

the analogous Friedmann equation

$$H^2 = -\frac{K}{a^2} + \frac{8\pi G\,\rho_{(s)}}{3\,a^6} + \frac{8\pi G\,\rho_{(r)}}{3\,a^4} \qquad (8.33)$$

describes a universe with negative curvature, zero cosmological constant, and filled with a stiff fluid and a radiation fluid. The latter, however, has negative energy density: more precisely,

$$K = -\left(\frac{\alpha}{\beta}\right)^2 = -\tan^2\vartheta\,, \qquad (8.34)$$

$$8\pi G\,\rho_{(s)} = \frac{3\,v_0^2}{\beta^2}\,, \qquad (8.35)$$

$$8\pi G\,\rho_{(r)} = -\frac{6\,\alpha\,v_0}{\beta^2} < 0\,. \qquad (8.36)$$

The Lagrangian for the analogous FLRW cosmos inspires the Lagrangian for the lava flow problem

$$L\,(h, h') = h(h')^2 - \frac{2\,\alpha\,v_0}{\beta^2 h} + \frac{v_0^2}{\beta^2\,h^3} + \left(\frac{\alpha}{\beta}\right)^2 h\,. \qquad (8.37)$$

Since $\partial L/\partial x = 0$, the corresponding Hamiltonian

$$\mathcal{H} = h(h')^2 + \frac{2\alpha\,v_0}{\beta^2 h} - \frac{v_0^2}{\beta^2 h^3} - \left(\frac{\alpha}{\beta}\right)^2 h \qquad (8.38)$$

is constant. Setting $\mathcal{H} = 0$, we recover Eq. (8.32) for the lava flow profile.
   The analytic solution of the lava problem [115]

$$x_0 - x = H_0 \cot\vartheta \left[\tanh^{-1}\left(\frac{h}{H_0}\right) - \frac{h}{H_0}\right]\,, \qquad (8.39)$$

where $0 \leq x \leq x_0$ and

$$H_0 = \sqrt{\frac{3\eta\,v_0}{\rho\,g\sin\vartheta}} = \sqrt{\frac{v_0}{\alpha}} = \sqrt{\frac{2\rho_{(s)}}{|\rho_{(r)}|}}\,, \qquad (8.40)$$

gives the solution of the analogous universe

$$\frac{t_0 - t}{\tau} = \tanh^{-1}\left(\frac{a}{a_0}\right) - \frac{a}{a_0} \qquad (8.41)$$

for $0 \leq t \leq t_0$, where

$$\tau = \sqrt{\frac{2\rho_{(s)}}{|\rho_{(r)}|}}\,\cot\vartheta\,. \qquad (8.42)$$

Since the analogous universe is spatially curved, the familiar symmetries of the Einstein–Friedmann equations do not apply.

By neglecting the slope and setting $\sin\vartheta \simeq 0$, $\cos\vartheta \simeq 1$ (which corresponds to zero spatial curvature in the cosmic analogy), Ref. [115] obtains the approximate profile

$$h(x) \simeq \left[\frac{9\,\eta\,v_0}{\rho\,g}\,(x_0 - x)\right]^{1/3} \tag{8.43}$$

for $0 \le x \le x_0$, which has been proposed to describe the shape of an accretionary wedge in the oceanic crust [128]. In this approximation, only the stiff fluid remains in the analogous Friedmann equation, the curvature disappears and the scale factor of the analogous spatially flat universe reduces to the well known power–law $a(t) \simeq a_0\,(t_0 - t)^{1/3}$.

8.5.3 *River floods*

The scientific study and mathematical modelling of river flooding began in the mid 1800s. Although there is a variety of flood phenomena, non–extreme floods in long rivers can be modelled as kinematic waves (Fig. 8.3).

Contrarily to dynamic waves, which are ultimately derived from Newton's second law, kinematic waves are derived from a conservation equation,

FIG. 8.3. Flood wave in the Isar river, Bayern, Germany.

for example the kinematic waves found in traffic flow, in glaciers, in river floods and in many other phenomena (see, *e.g.*, [169,237,263,293,318,323]).

Consider a long straight river on a flat bed of slope S_0, let x be a coordinate along the flow, $h(x)$ be the depth of the water, and v its one–dimensional velocity. Considering a static profile with constant depth $h = h_0$, constant velocity $v = v_0$, and small amplitude kinematic waves around it, as described by the expansion

$$h(t) = h_0 + \tau\, h_1(t) + \tau^2\, h_2(t) + \dots, \tag{8.44}$$

$$v(t) = v_0 + \tau\, v_1(t) + \tau^2\, v_2(t) + \dots, \tag{8.45}$$

Lighthill & Whitham [262] obtain the first order equation for the propagation of the perturbations

$$\frac{dh_1}{dt} = \frac{3}{2h_0\,(1+F)}\,h_1^2 - \frac{g\,S_0}{v_0}\left(1 - \frac{F}{2}\right)h_1\,, \tag{8.46}$$

where g is the acceleration of gravity and

$$F = \frac{v_0}{\sqrt{g\,h_0}} \tag{8.47}$$

is the Froude number. This equation has the form

$$\dot h_1 = \alpha\, h_1^2 - \beta\, h_1\,. \tag{8.48}$$

Since only $F < 1$ is physical [262], introducing the constants

$$b = g S_0 \left(1 - \frac{F}{2}\right), \tag{8.49}$$

$$K = \frac{g\,h_0\,S_0}{3v_0}\,(2-F)\,(1+F) = \frac{2b\,h_0}{3v_0}\,(1+F)\,, \tag{8.50}$$

the solution corresponding to the initial condition $h_1(0)$ at time $t = 0$ is [262]

$$h_1(t) = \frac{K\,h_1(0)\,e^{-bt}}{K - h_1(0)\left(1 - e^{-bt}\right)}. \tag{8.51}$$

If we introduce $f(t) \equiv \dfrac{\alpha}{\beta}\,h_1(t)$, then Eq. (8.48) becomes $\dot f = \beta f\left(f - 1\right)$, in which the right hand side has opposite sign to the right hand side of the logistic equation (2.85), so Eq. (8.48) describes decay instead of growth, corresponding to the fact that a flood wave decreases in time. The variable change

$$f = 1 - \psi \tag{8.52}$$

gives the logistic equation $\dot{\psi} = \beta \psi (1 - \psi)$ for the new variable ψ and its sigmoid solution (2.86) then reproduces the Lighthill–Whitham solution (8.51).

Dividing Eq. (8.48) by h_1 and squaring both sides, one obtains

$$\left(\frac{\dot{h}_1}{h_1} \right)^2 = \alpha^2 h_1^2 - 2\alpha \beta h_1 + \beta^2 \, ; \tag{8.53}$$

the analogous Friedmann equation

$$H^2 = \frac{8\pi G}{3} \rho_1 a^2 + \frac{8\pi G}{3} \rho_2 a + \frac{\Lambda}{3} \tag{8.54}$$

is obtained from the correspondence

$$\Lambda = 3\beta^2 \, , \tag{8.55}$$

$$\frac{8\pi G}{3} \rho_1 = \alpha^2 > 0 \, , \tag{8.56}$$

$$\frac{8\pi G}{3} \rho_2 = -2\alpha\beta < 0 \, , \tag{8.57}$$

and is spatially flat, has positive cosmological constant and contains two phantom fluids with equation of state parameters $w_1 = -5/3$ and $w_2 = -4/3$. The latter has negative energy density. The effective Lagrangian and Hamiltonian and the symmetries can be obtained from those of the logistic equation discussed in Chap. 2.

8.6 Conclusions

The long list of possible analogies between cosmology and various physical systems is not finished, but this book must come to an end. Our goal was not to provide an exhaustive list of such systems anyway, or even a list of analogous natural systems and phenomena (on which we have put some emphasis), but to show that these analogies exist, that they can be intriguing, and that sometimes they are useful to develop new knowledge or to improve the existing understanding.

At this point, one must ask oneself: Are analogies useful? What have we learned thus far? A gut reaction could be that this question may have to be answered on a case–by–case basis. For example, in the analogies presented in the previous chapters we have found a new invariant of the Fibonacci sequence, which may not be of practical importance but is nevertheless a little step forward in mathematics that has been around since the middle ages. The cosmological analogy is definitely useful in the study of equilibrium beach profiles, helping us to reach definite conclusions in a debate about the analytical solutions of a theory of the generation of these beach profiles. Furthermore, we have found that, in this theory, equilibrium beach profiles correspond to minimum dissipation rates at the sea bottom.

The cosmic analogy is useful to understand better the Vialov equation ruling the longitudinal profiles (the "shape") of glaciers and ice caps. The mathematical physics underlying longitudinal glacier profiles emerges with a bit more clarity when comparing the Vialov equation with its cosmological cousin, leading to a generating function and to a formal representation of the solutions of the Vialov equation.

Another cosmic analogy is useful in the description of the transverse profiles of glacial valleys and augments the catalogue of analytical solutions of this earth science problem. The parallel between equilibrium beach profiles and glacial valley configurations cannot escape the reader's attention and leads also to confirming that the latter maximize the friction between the glacier and the valley walls and bed. It is refreshing to unearth similarities in fields apparently so distant from each other as oceanography and glaciology. Incidentally, glacial valley profiles could be called "equilibrium valley profiles", borrowing from the terminology of coastal morphology (the concept of equilibrium is implicit in the glaciology context but parallel terminology in the earth sciences, which use vastly different nomenclatures, would be nice for a unified view and would certainly benefit the student). To continue, the old and already solved terrestrial brachistochrone problem allows one to write immediately the solution for a universe with a complicated (phantom and non-linear) equation of state, and the list goes on.

The next thought that comes to mind when attempting to answer the question of whether analogies are useful brings up the unifying theme of the inverse variational problem, *i.e.*, of finding Lagrangians and Hamiltonians for the differential equations describing the systems analogous to FLRW universes. The inverse variational problem is definitely of interest in pure and applied mathematics and appears in the description of many physical systems or phenomena through ordinary differential equations. These equations are often derived from a purely phenomenological point of view in which a Lagrangian or a Hamiltonian formulation were far from the intentions of the scientists proposing a model, and were not contemplated in previous literature. Through cosmological analogies, the Lagrangian and Hamiltonian formulations of FLRW cosmology suggest the solutions to many inverse variational problems, for example providing Lagrangians and Hamiltonians for heating and cooling, landslides and debris flows, the longitudinal profiles of glaciers, Omori's law for earthquake aftershocks, lawa flows, *etc.* As a second unifying theme, a symmetry of the Einstein–Friedmann equations of cosmology for spatially flat universes dominated by a single perfect fluid can sometimes unveil a previously unknown symmetry of the equations ruling an analogous physical system. Another unifying theme is that the graphs of the solutions of all systems admitting an analogy with FLRW cosmology are roulettes.

On a deeper level, answering the practical question of whether analogies are useful leads one to question again the relevance of usefulness in pure science. Most of pure science is driven by sheer curiosity and the need to understand nature, not by questions of immediate practical interest (although, eventually, new discoveries and new understanding always find their way in more practical formulations or applications). Analogies suggest intriguing ideas that, although without use in the short term, broaden our view of certain phenomena or our global view of science and of the natural world. For example, when discussing the analogy between freezing lakes and FLRW cosmology, we have mused about ideas on spacetime in the early universe near the Big Bang or other cosmological singularities, imagining in loose terms the possibility of another "phase" of gravity or of spacetime which presumably is, or will be, addressed by quantum gravity and quantum cosmology. The analogy between the hypothetical cosmic Big Rip and the main shock of an earthquake is also intriguing, leading one to notice common features in destructive or runaway phenomena from both the mathematical and the physical points of view. The value of these analogies consists of stimulating the imagination and of generating mental images and ideas that may promote new directions in research. The details and limits of applicability of any specific analogy are of secondary importance. If this book succeeds in stimulating the imagination of the reader, it will have achieved its goal.

Appendix A

A.1 Einstein equations for a spatially flat FLRW universe

Here we compute explicitly the Ricci tensor and the Einstein equations for a spatially flat universe.

The FLRW line element (1.127) contains only one unknown function, the scale factor $a(t)$. Because of the high degree of symmetry, the Einstein partial differential equations reduce to ordinary differential equations in FLRW cosmology. If the cosmological constant Λ is non–zero, we incorporate it into the effective stress–energy tensor $T_{\mu\nu}^{(\Lambda)} = -\dfrac{\Lambda}{8\pi G}\, g_{\mu\nu}$, moving it to the right hand side of the Einstein equations (1.106), which become

$$\mathcal{R}_{\mu\nu} - \frac{1}{2} g_{\mu\nu}\mathcal{R} = 8\pi G\, T_{\mu\nu}^{(\text{total})}, \tag{A.1}$$

where

$$T_{\mu\nu}^{(\text{total})} = T_{\mu\nu} + T_{\mu\nu}^{(\Lambda)}. \tag{A.2}$$

In Cartesian comoving coordinates $\big(t, x, y, z\big)$, the components of the metric and inverse metric tensors of a spatially flat ($\kappa = 0$) FLRW universe are given by the matrices

$$\big(g_{\mu\nu}\big) = \begin{pmatrix} -1 & 0 & 0 & 0 \\ 0 & a^2 & 0 & 0 \\ 0 & 0 & a^2 & 0 \\ 0 & 0 & 0 & a^2 \end{pmatrix}, \tag{A.3}$$

$$\big(g^{\mu\nu}\big) = \begin{pmatrix} -1 & 0 & 0 & 0 \\ 0 & 1/a^2 & 0 & 0 \\ 0 & 0 & 1/a^2 & 0 \\ 0 & 0 & 0 & 1/a^2 \end{pmatrix}. \tag{A.4}$$

223

Let us compute the Christoffel symbols

$$\Gamma^{\rho}_{\mu\nu} = \frac{1}{2} g^{\rho\sigma} \left(g_{\sigma\mu,\nu} + g_{\sigma\nu,\mu} - g_{\mu\nu,\sigma} \right). \tag{A.5}$$

- For $\mu = i, \nu = j$ it is

$$\Gamma^{\rho}_{ij} = \frac{1}{2} g^{\rho\sigma} \left(g_{i\sigma,j} + g_{j\sigma,i} - g_{ij,\sigma} \right)$$

$$= -\frac{1}{2} g^{\rho 0} g_{ij,0} = -\frac{1}{2} \delta^{\rho 0} g^{00} g_{ii,0} \delta_{ij} = \delta^{\rho 0} \delta_{ij} \, a \, \dot{a}.$$

- For $\mu = i, \nu = 0$ it is

$$\Gamma^{\rho}_{i0} = \frac{1}{2} g^{\rho\sigma} \left(g_{i\sigma,0} + g_{0\sigma,i} - g_{i0,\sigma} \right)$$

$$= \frac{1}{2} g^{\rho i} g_{ii,0} = -\frac{1}{2} \delta^{\rho}_{i} g^{kk} g_{kk,0} = \frac{1}{2} \delta^{\rho}_{i} \frac{2 \, a \, \dot{a}}{a^2} = \delta^{\rho}_{i} \frac{\dot{a}}{a}.$$

- For $\mu = \nu = 0$ it is

$$\Gamma^{\rho}_{00} = \frac{1}{2} g^{\rho\sigma} \left(2 \, g_{0\sigma,0} - g_{00,\sigma} \right) = g^{\rho 0} \, g_{00,0} = 0. \tag{A.6}$$

Putting everything together, the only non-vanishing Christoffel symbols are

$$\Gamma^{0}_{11} = \Gamma^{0}_{22} = \Gamma^{0}_{33} = a \, \dot{a}, \tag{A.7}$$

$$\Gamma^{1}_{01} = \Gamma^{1}_{10} = \Gamma^{2}_{02} = \Gamma^{2}_{20} = \Gamma^{3}_{03} = \Gamma^{3}_{30} = \frac{\dot{a}}{a}. \tag{A.8}$$

The components of the Ricci tensor in comoving coordinates are then calculated using

$$\mathcal{R}_{\mu\nu} = \Gamma^{\nu}_{\mu\rho,\nu} - \Gamma^{\nu}_{\nu\rho,\mu} + \Gamma^{\alpha}_{\mu\rho} \Gamma^{\nu}_{\alpha\nu} - \Gamma^{\alpha}_{\nu\rho} \Gamma^{\nu}_{\alpha\mu}. \tag{A.9}$$

Due to spatial isotropy, all the off–diagonal components vanish and we are left with

$$\mathcal{R}_{00} = \Gamma^\nu_{00,\nu} - \Gamma^\nu_{\nu 0,0} + \Gamma^\alpha_{00}\Gamma^\nu_{\alpha\nu} - \Gamma^\alpha_{\nu 0}\Gamma^\nu_{\alpha 0}$$

$$= -\left(\Gamma^1_{10,0} + \Gamma^2_{20,0} + \Gamma^3_{30,0}\right) - \left(\Gamma^1_{\nu 0}\Gamma^\nu_{10} + \Gamma^2_{\nu 0}\Gamma^\nu_{20} + \Gamma^3_{\nu 0}\Gamma^\nu_{30}\right)$$

$$= -3\left(\frac{\dot{a}}{a}\right)_{,0} - \left[\left(\Gamma^1_{10}\right)^2 + \left(\Gamma^2_{20}\right)^2 + \left(\Gamma^3_{30}\right)^2\right]$$

$$= \frac{3\,\dot{a}^2}{a^2} - \frac{3\,\ddot{a}}{a} - \frac{3\,\dot{a}^2}{a^2} = -\frac{3\,\ddot{a}}{a}\,,$$

$$\mathcal{R}_{11} = \Gamma^\nu_{11,\nu} - \Gamma^\nu_{\nu 1,1} + \Gamma^\alpha_{11}\Gamma^\nu_{\alpha\nu} - \Gamma^\alpha_{\nu 1}\Gamma^\nu_{\alpha 1}$$

$$= \Gamma^0_{11,0} + \Gamma^0_{11}\Gamma^\nu_{0\nu} - \left(\Gamma^0_{\nu 1}\Gamma^\nu_{01} + \Gamma^1_{\nu 1}\Gamma^\nu_{11} + \Gamma^3_{\nu 1}\Gamma^\nu_{31}\right)$$

$$= \Gamma^0_{11,0} + \Gamma^0_{11}\left(\Gamma^1_{01} + \Gamma^2_{02} + \Gamma^3_{03}\right) - \left(\Gamma^0_{11}\Gamma^1_{01} + \Gamma^1_{01}\Gamma^0_{11}\right)$$

$$= (a\,\dot{a})_{,0} + a\,\dot{a}\,\frac{3\,\dot{a}}{a} - 2\,\dot{a}^2 = 2\,a\dot{a}^2 + a\,\ddot{a}\,.$$

Due to spatial isotropy, it is

$$\mathcal{R}_{11} = \mathcal{R}_{22} = \mathcal{R}_{33}\,, \tag{A.10}$$

as is straightforward to verify. The Ricci scalar is then

$$\mathcal{R} = g^{00}\,\mathcal{R}_{00} + 3\,g^{11}\,\mathcal{R}_{11} = 6\left(\dot{H} + 2H^2\right)\,. \tag{A.11}$$

The $(0,0)$ component of the Einstein tensor is

$$G_{00} = \mathcal{R}_{00} - \frac{1}{2}\,g_{00}\,\mathcal{R} = -\frac{3\,\ddot{a}}{a} + \frac{1}{2}\cdot 6\left(\dot{H} + 2H^2\right) = 3H^2 \tag{A.12}$$

and the $(0,0)$ component of the Einstein equations (A.1) gives the Friedmann equation

$$H^2 = \frac{8\pi G}{3}\,\rho\,. \tag{A.13}$$

The $(1,1)$ component of the Einstein tensor is

$$G_{11} = \mathcal{R}_{11} - \frac{1}{2} g_{11} \, \mathcal{R} = 2\,\dot{a}^2 + a\,\ddot{a} - \frac{a^2}{2} \cdot 6 \left(\dot{H} + 2H^2 \right) = -\dot{a}^2 - 2\,a\,\ddot{a} \quad \text{(A.14)}$$

and the Einstein equations (A.1) give

$$-\frac{2\,\ddot{a}}{a} - \frac{\dot{a}^2}{a^2} \, \rho = 8\pi G \, P \, . \qquad\qquad \text{(A.15)}$$

The $(2,2)$ and $(3,3)$ components of the Einstein equations coincide with the $(1,1)$ component and do not provide additional information.

Appendix B

B.1 Analogue of the terrestrial brachistochrone

Here we present a stability analysis of the static universe analogous to the terrestrial brachistochrone. Consider a perturbation of this static universe solution such that

$$a(t) = a_{\min} + \delta a(t) , \tag{B.1}$$

where $\delta a \geq 0$ since it must be $a_{\min} \leq a(t) < a_0$. The zero order acceleration equation (1.180) gives

$$1 - \frac{4\pi G}{3} (\rho_{\min} + 3P_{\min}) = 0 , \tag{B.2}$$

where $\rho_{\min}$ and $P_{\min}$ are the energy density and pressure of the unperturbed static universe with scale factor $a = a_{\min}$.

The acceleration equation (1.180) for our situation can be written as

$$\ddot{a} = \frac{a^3 \left(2a_0^2 - a^2 \right)}{C^2 \left(a_0^2 - a_{\min}^2 \right)^2} + a . \tag{B.3}$$

We now expand this acceleration equation to first order in the perturbation $\delta a / a_{\min} \ll 1$ and use Eq. (4.56). Retaining only the zero order and the first order terms, one obtains

$$\delta \ddot{a} \simeq \frac{C^2}{a_{\min}} \left\{ 2a_0^2 - a_{\min}^2 + \left[\frac{2C^2}{a_{\min}} \left(2a_0^2 - a_{\min}^2 + 1 \right) + \frac{1}{a_{\min}} + 1 \right] \delta a \right\} + a_{\min} . \tag{B.4}$$

Substituting the zero order equation (B.2) yields the first order acceleration equation satisfied by the perturbations δa,

$$\delta \ddot{a} = \Omega^2 \, \delta a , \tag{B.5}$$

where the angular frequency squared

$$\Omega^2 = 1 + \frac{C^2}{a_{\min}^2} \left[2C^2 \left(1 + 2a_0^2 - a_{\min}^2 \right) + 1 \right] \qquad \text{(B.6)}$$

is always positive because $a_0 > a_{\min}$. Finally, the linear order solution is

$$\delta a(t) = A \, e^{\Omega t} + B \, e^{-\Omega t} \qquad \text{(B.7)}$$

with $\Omega \in \mathbb{R}$ and where A and B are arbitrary constants. This perturbation includes a mode that diverges exponentially fast with time and the static universe is clearly unstable.

Appendix C

C.1 Radial geodesics of static spherical geometries

Let us examine examples of cosmic analogies that originate from the equation of radial geodesics in static and spherically symmetric metrics. We begin with situations in which $g_{tt}\, g_{RR} = -1$ and $\Phi \equiv 0$.

C.1.1 *Reissner–Nordström black hole*

An example is the Reissner–Nordström spacetime with line element [82, 120, 202, 435]

$$ds^2 = -\left(1 - \frac{2m}{R} + \frac{Q^2}{R^2}\right) dt^2 + \frac{dR^2}{1 - \frac{2m}{R} + \frac{Q^2}{R^2}} + R^2 d\Omega_{(2)}^2 \,, \tag{C.1}$$

where m and Q are the mass and charge parameters. The Misner–Sharp–Hernandez mass [205, 206, 283] is

$$M(R) = m - \frac{Q^2}{2R} \tag{C.2}$$

and the energy density of the cosmic fluid in the analogous FLRW cosmos is

$$\rho(a) = \frac{3}{4\pi}\left(\frac{m}{a^3} - \frac{Q^2}{2\,a^4}\right) \,. \tag{C.3}$$

The second term in the right-hand side describes a radiation fluid with negative energy density, which is unphysical. However, when $Q = 0$ and Reissner–Nordström reduces to Schwarzschild, the energy density is just the first term $\rho = \rho_0/a^3$ describing a dust (for any possible value of K), as already seen.

C.1.2   *(Anti–)de Sitter universe*

Another example is (Anti–)de Sitter space, the geometry of which is described by

$$ds^2 = -\left(1 \mp H^2 R^2\right) dt^2 + \frac{dR^2}{1 \mp H^2 R^2} + R^2 d\Omega_{(2)}^2 \qquad (\text{C.4})$$

in locally static coordinates, where the upper [lower] sign refers to de Sitter [Anti–de Sitter] space. The corresponding Misner–Sharp–Hernandez mass is

$$M(R) = \pm \frac{H^2 R^3}{2} \qquad (\text{C.5})$$

and the analogous universe is filled by a fluid with energy density

$$\rho(a) = \pm \frac{3H^2}{8\pi} \qquad (\text{C.6})$$

and pressure

$$P = -\frac{M'}{4\pi a^2} = \mp \frac{3H^2 a^2}{8\pi a^2} = -\rho, \qquad (\text{C.7})$$

which describe the cosmological constant $\Lambda = \pm 3\,H^2$. The timelike geodesics of (Anti–)de Sitter produce analogous universes with the same cosmological constant and any value of the curvature index. In particular, if $\kappa = 0, -1$, the analogy reproduces the same (Anti–)de Sitter space used to generate analogies.

C.1.3   *Schwarzschild–de Sitter/Kottler spacetime*

Next, we examine the Schwarzschild–de Sitter/Kottler geometry

$$ds^2 = -\left(1 - \frac{2m}{R} - H^2 R^2\right) dt^2 + \frac{dR^2}{1 - \frac{2m}{R} - H^2 R^2}$$

$$+ R^2 d\Omega_{(2)}^2 , \qquad (\text{C.8})$$

which has Misner–Sharp–Hernandez mass

$$M = m + \frac{H^2 R^3}{2} . \qquad (\text{C.9})$$

The analogous FLRW universe has energy density

$$\rho(a) = \frac{3m}{4\pi}\left(\frac{1}{a^3} + \frac{H^2}{2}\right) , \qquad (\text{C.10})$$

which describes a dust plus Λ. Given that the Schwarzschild black hole generates a dust–dominated analogous FLRW universe and de Sitter space has itself as an analogue, this conclusion is not surprising.

C.1.4 *Kiselev black hole*

The Kiselev line element [240] describes a black hole embedded in a mixture of fluids with anisotropic pressure (contrary to appearances, it is not asymptotically FLRW nor a perfect fluid solution of the Einstein equations (1.106) [424]), and it reads

$$ds^2 = -f(R)\, dt^2 + \frac{dR^2}{f(R)} + R^2 d\Omega^2_{(2)}, \tag{C.11}$$

where

$$f(R) = 1 - \frac{2m}{R} - \sum_n \left(\frac{r_n}{R}\right)^{3w_n+1}, \tag{C.12}$$

where m, r_n, and w_n are constants and $-1 < w_n < -1/3$ [240]. Here the Misner–Sharp–Hernandez mass is

$$M(R) = m + \frac{R}{2} \sum_n \left(\frac{r_n}{R}\right)^{3w_n+1}, \tag{C.13}$$

producing the energy density and pressure of the analogous FLRW universe

$$\rho = \rho_{\text{dust}} + \sum_n \frac{3\, a_n^{3w_n+1}}{8\pi} \frac{1}{a^{3(w_n+1)}}, \tag{C.14}$$

$$P = \sum_n w_n\, \rho_n, \tag{C.15}$$

where

$$\rho_{\text{dust}} = \frac{\rho_0}{a^3}, \tag{C.16}$$

$$\rho_0 = \frac{3}{4\pi}, \tag{C.17}$$

$$\rho_n - \frac{3\, a_n^{3w_n+1}}{8\pi}. \tag{C.18}$$

In this analogy, the Kiselev anisotropic fluid becomes a mixture of (isotropic) perfect dark energy fluids (see [157] for a review of two- and three-fluid analytical solutions of the Einstein-Friedmann equations).

C.1.5 *Barriola–Vilenkin global monopole*

The Barriola–Vilenkin global monopole (or "string hedgehog") has the geometry [24, 189]

$$ds^2 = -\left(1 - 8\pi\eta^2 - \frac{2m}{R}\right) dt^2 + \frac{dr^2}{1 - 8\pi\eta^2 - 2m/R} + R^2 d\Omega^2_{(2)} \tag{C.19}$$

with constant m, η and the Misner–Sharp–Hernandez mass is $M(R) = m + 4\pi\eta^2 R$, producing the energy density

$$\rho(a) = \frac{3\,m}{4\pi a^3} + \frac{3\,\eta^2}{a^2} \tag{C.20}$$

and the analogous Friedmann equation

$$H^2 = \frac{(\bar{E}^2 - 1 + 3\,\eta^2)}{a^2} + \frac{\rho_0}{a^3} \tag{C.21}$$

with

$$K = 1 - \bar{E}^2 - 3\,\eta^2\,, \tag{C.22}$$

$$\rho_0 = \frac{3\,m}{4\pi}\,. \tag{C.23}$$

This is a spatially curved universe (except for special values of $\bar{E}$ and η) filled with dust.

C.1.6 *Bardeen regular black hole*

The Bardeen regular black hole [22] quantum–corrects the Schwarzschild black hole to remove the central singularity and it solves the Einstein equations coupled to non-linear electrodynamics [14]. The line element

$$ds^2 = -\left[1 - \frac{2mR^2}{(R^2 + a^2)^{3/2}}\right] dt^2 + \frac{dR^2}{1 - \frac{2mR^2}{(R^2 + a^2)^{3/2}}} + R^2 d\Omega_{(2)}^2 \tag{C.24}$$

gives the Misner–Sharp–Hernandez mass

$$M(R) = \frac{m\,R^3}{(R^2 + a^2)^{3/2}} \tag{C.25}$$

corresponding to the energy density and pressure of the analogous FLRW universe

$$\rho(a) = \frac{3\,m}{4\pi} \frac{1}{\left(a^2 + a_0^2\right)^{3/2}}\,, \tag{C.26}$$

$$P = -\left(\frac{4\pi}{3m}\right)^{2/3} a^2 \rho^{5/3}\,, \tag{C.27}$$

satisfying a phantom and non-linear equation of state.

Let us move now to static spherical geometries in which $\Phi \neq 0$.

C.1.7 *Morris–Thorne wormhole*

Let us consider the Morris–Thorme wormhole [287] with line element

$$ds^2 = -dt^2 + dr^2 + \left(b_0^2 + r^2\right)d\Omega_{(2)}^2, \tag{C.28}$$

where b_0 is a constant and the areal radius is $R = \sqrt{b_0^2 + r^2}$. Substituting $dr = R\,dR/r$ into Eq. (C.28) gives

$$ds^2 = -dt^2 + \frac{R^2}{R^2 - b_0^2}\,dR^2 + R^2 d\Omega_{(2)}^2. \tag{C.29}$$

The Misner–Sharp–Hernandez mass [205, 206, 283] is $M(R) = b_0^2/(2R)$, while

$$e^{-2\Phi} = \frac{1}{1 - 2m/R} \tag{C.30}$$

and the energy density of the FLRW cosmic analogue is

$$\rho(a) = \frac{3\,b_0^2}{8\pi\,a^4}. \tag{C.31}$$

The Friedmann equation satisfied by the analogous FLRW universe reads

$$H^2 = \frac{\left(E^2 - 1\right)}{a^2} + \frac{b_0^2\left(1 - E^2\right)}{a^4}, \tag{C.32}$$

which makes sense physically if $E^2 < 1$ and negative energy densities are avoided; then the analogous universe is spatially curved and contains a radiation fluid.

Radial null geodesics produce the analogous Friedmann equation

$$H^2 = \frac{E^2}{a^2} - \frac{b_0^2\,E^2}{a^3} \tag{C.33}$$

with a dust of negative energy density, which is unphysical.

C.1.8 *Wyman's "other" solution*

Wyman's "other" solution [446] is a little known scalar field solution of the Einstein equations, not to be confused with the better known solution discovered by Fisher and rediscovered by Bergmann & Leipnik, Janis, Newman & Winicour, Buchdahl, and Wyman [38, 38, 66, 165, 232, 422, 446]. The line element is

$$ds^2 = -R^2 dt^2 + 2\,dR^2 + R^2 d\Omega_{(2)}^2 \tag{C.34}$$

with time-dependent scalar field source $\phi(t) = \phi_0\, t$. Here

$$e^{-2\Phi} = \frac{R^2}{1 - 2M/R}, \tag{C.35}$$

$M(R) = R/4$, and the analogous Friedmann equation is

$$H^2 = \frac{8\pi\rho_0}{3a^4} - \frac{K}{a^2} \tag{C.36}$$

where $K = -1/2$. This analogous universe is positively curved and filled with radiation.

Radial null geodesics, instead, produce the analogous Friedmann equation $H^2 = E^2/(2\,a^4)$ describing a spatially flat, radiation–dominated FLRW universe.

The geometry (C.34) has been generalized by including a cosmological constant Λ [389] and studied in [17, 158]:

$$ds^2 = -R^2 dt^2 + \frac{2\, dR^2}{1 - 2\Lambda R^2/3} + R^2 d\Omega^2_{(2)}. \tag{C.37}$$

Looking at radial timelike geodesics, the Misner–Sharp–Hernandez mass is now

$$M(R) = \frac{R}{4} + \frac{\Lambda R^3}{6} \tag{C.38}$$

and the analogous Friedmann equation becomes

$$H^2 = \frac{\bar{E}^2\,(1 + \Lambda/3) - 1/2}{a^2} + \frac{\Lambda}{3} + \frac{\bar{E}^2}{2a^4}, \tag{C.39}$$

corresponding to a FLRW universe with curvature index

$$K = \frac{1}{2} - \bar{E}^2\left(1 + \frac{\Lambda}{3}\right), \tag{C.40}$$

cosmological constant Λ, and a radiation fluid with $\rho_0 = \bar{E}^2/2$. The analogy stemming from radial null geodesics is unchanged.

Bibliography

[1] G. Abreu and M. Visser, Kodama time: Geometrically preferred foliations of spherically symmetric spacetimes, *Phys. Rev. D* **82**, 044027 (2010).

[2] I. Affleck, Golden mean seen in a magnet, *Nature* **464**, 362 (2010).

[3] J. Alberto Vazquez, S. Hee, M. P. Hobson, A. N. Lasenby, M. Ibison, and M. Bridges, Observational constraints on conformal time symmetry, missing matter and double dark energy, *J. Cosmol. Astropart. Phys.* **07**, 062 (2018).

[4] H. Aldersey–Williams, *The Most Beautiful Molecule: The Discovery of the Buckyball*, Wiley, New York (1995).

[5] L. Amendola and S. Tsujikawa, *Dark Energy, Theory and Observations*, Cambridge University Press, Cambridge, UK (2010).

[6] K. N. Ananda and M. Bruni, Cosmodynamics and dark energy with non–linear equation of state: a quadratic model, *Phys. Rev. D* **74**, 023523 (2006).

[7] K. N. Ananda and M. Bruni, Cosmodynamics and dark energy with a quadratic EoS: anisotropic models, large–scale perturbations and cosmological singularities, *Phys. Rev. D* **74**, 023524 (2006).

[8] R. Antonelli and A. R. Klotz, A smooth trip to Alpha Centauri: Comment on "The least uncomfortable journey from A to B", *Am. J. Phys.* **85**, 469 (2017).

[9] I. Ya. Aref'eva, A. S. Koshelev, and S. Yu. Vernov, Exact solution in a string cosmological model, *Theor. Math. Phys.* **148**, 895 (2006).

[10] G. D. Ashton, River ice, *Am. Sci.* **67**, 38–45 (1979).

[11] G. D. Ashton, Thin ice growth, *Water Resour. Res.* **25/3**, 564–566 (1989).

[12] P. J. Aubusson, A. G. Harrison, and S. M. Ritchie (editors), *Metaphor and Analogy in Science Education*, Springer, Dordrecht (2006).

[13] A. Aufdermaur and D. A. Johnson, Charge separation due to riming in an electric field, *Quart. J. Roy. Meteorol. Soc.* **98**, 369 (1972).

[14] F. Ayón–Beato and A. García, The Bardeen model as a nonlinear magnetic monopole, *Phys. Lett. B* **493**, 149–152 (2000).

[15] D. B. Bahr, W. T. Pfeffer, and G. Kaser, A review of volume–area scaling of glaciers, *Rev. Geophys.* **10**, 95–140 (2015).

[16] K. Bamba, S. Nojiri, and S. D. Odintsov, The Universe future in modified gravity theories: Approaching the finite–time future singularity, *J. Cosmol. Astropart. Phys.* **10**, 045 (2008).

[17] A. Banijamali, B. Fazlpour, and V. Faraoni, Wyman's "other" scalar field solution, Sultana's generalization, and their Brans–Dicke and $\mathcal{R}^2$ relatives, *Phys. Rev. D* **100**, 064017 (2019).

[18] N. Barbosa–Cendejas and M. A. Reyes, The Schrödinger picture of standard cosmology, arXiv:1001.0084 [gr–qc].

[19] C. Barceló, S. Liberati, and M. Visser, Analog gravity from Bose–Einstein condensates, *Class. Quantum Grav.* **18**, 1137 (2001).

[20] C. Barceló, S. Liberati, and M. Visser, Analog models for FRW cosmologies, *Int. J. Mod. Phys. D* **12**, 1641 (2003).

[21] C. Barceló, M. Visser, and S. Liberati, Einstein gravity as an emergent phenomenon?, *Int. J. Mod. Phys. D* **10**, 799–806 (2001).

[22] J. M. Bardeen, Non–singular general–relativistic gravitational collapse, in *Proceedings of International Conference GR5*, Tbilisi, USSR 1968, p. 174.

[23] L. Baringhaus, Fibonacci numbers, Lucas numbers and integrals of certain Gaussian processes, *Proc. Am. Math. Soc.* **124**, 3875 (1996).

[24] M. Barriola and A. Vilenkin, Gravitational field of a global monopole, *Phys. Rev. Lett.* **63**, 341 (1989).

[25] J. D. Barrow, Relativistic cosmology and the regularization of orbits, *The Observatory* **113**, 210 (1993).

[26] J. D. Barrow, Sudden future singularities, *Class. Quantum Grav.* **21**, L79 (2004).

[27] J. D. Barrow, *The Book of Universes*, W. W. Norton & C., New York (2011).

[28] J. D. Barrow and A. Paliathanasis, Reconstructions of the dark–energy equation of state and the inflationary potential, *Gen. Relativ. Gravit.* **50**, 82 (2018).

[29] J. D. Barrow, G. Galloway, and F. J. T. Tipler, The closed–universe recollapse conjecture, *Mon. Not. Roy. Astr. Soc.* **223**, 835 (1986).

[30] J. D. Barrow, G. F. R. Ellis, R. Maartens, and C. G. Tsagas, On the stability of the Einstein static universe, *Class. Quantum Grav.* **20**, L155–L164 (2003).

[31] W. Bascom, *Waves and Beaches, The Dynamics of the Ocean Surface*, Anchor Books, New York (1980).

[32] S. Basilakos, M. Tsamparlis, and A. Paliathanasis, Using the Noether symmetry approach to probe the nature of dark energy, *Phys. Rev. D* **83**, 103512 (2011).

[33] T. Battefeld and S. Watson, String gas cosmology, *Rev. Mod. Phys.* **78**, 435 (2006).

[34] R. A. Battye, M. Bucher, and D. Spergel, Domain wall dominated universes, arXiv:astro-ph/9908047.

[35] F. D. Belgiorno, S. L. Cacciatori, and D. Faccio, *Hawking Radiation: From Astrophysical Black Holes to Analogous Systems in The Lab*, World Scientific, Singapore (2019).

[36] J. Beltrán Jiménez, D. Rubiera-Garcia, D. Sáez–Gómez, and V. Salzano, Q–singularities, *Phys. Rev. D* **94**, 123520 (2016).

[37] D. I. Benn and N. R. J. Hulton, An excel spreadsheet program for reconstructing the surface profile of former mountain glaciers and ice caps, *Comput. Geosci.* **36**, 605–610 (2010).

[38] O. Bergmann and R. Leipnik, Space–Time Structure of a Static Spherically Symmetric Scalar Field, *Phys. Rev.* **107**, 1157 (1957).

[39] J. Bernstein and G. Feinberg (editors), *Cosmological Constants: Papers in Modern Cosmology*, Columbia University Press, New York (1986).

[40] E. Bertschinger, Cosmological dynamics, arXiv:astro-ph/9503125.

[41] D. Bini and S. Succi, Analogy between capillary motion and Friedmann–Robertson–Walker cosmology, *Europhys. Lett.* **82**, 34003 (2008).

[42] M. L. Boas, *Mathematical Methods in the Physical Sciences*, Wiley, Hoboken, NJ (1966).

[43] I. Bochicchio and E. Laserra, On the mechanical analogy between the relativistic evolution of a spherical dust universe and the classical motion of falling bodies, *J. Interdiscipl. Math.* **10**, 747–755 (2007).

[44] I. Bochicchio, S. Capozziello, and E. Laserra, The Weierstrass criterion and the Lemaître–Tolman–Bondi models with cosmological constant Λ, *Int. J. Geom. Meth. Mod. Phys.* **8**, 1653–1666 (2011).

[45] K. R. Bodge, Representing equilibrium beach profiles with an exponential expression, *J. Coastal Res.* **8**, 47–55 (1992).

[46] G. Böðvardsson, On the flow of ice–sheets and glaciers, *Jökull* **5**, 1–8 (1955).

[47] E. Boeker and R. van Grondelle, *Environmental Physics*, J. Wiley & Sons, Chicester (1999).

[48] K. Bohlin, Note sur le problème des deux corps et sur une intégration nouvelle dans le problème des trois corps, *Bull. Astron.* **28**, 113–119 (1911).

[49] C. F. Bohren, The freezing of streams and ponds: a simple–but uncomfortable–experiment, *Phys. Teach.* **42**, 522–525 (2004).

[50] B. M. Boman, T. N. Dinh, K. Decker, B. Emerick, C. Raymond, and G. Schleiniger, Why do Fibonacci numbers appear in patterns of growth in nature? A model for tissue renewal based on asymmetric cell division, *Fibonacci Quart.* **55**, 30–41 (2017).

[51] B. M. Boman, Y. Ye, K. Decker, C. Raymond, and G. Schleiniger, Geometric Branching Patterns based on p–Fibonacci Sequences: Self-similarity Across Different Degrees of Branching and Multiple Dimensions, *Fibonacci Quart.* **7**, 29–41 (2019).

[52] V. R. Bond, A transformation of the two–body problem, *Celestial Mechanics* **35**, 1–7 (1985).

[53] H. Bondi, Gravitational bounce in general relativity, *Mon. Not. Roy. Astr. Soc.* **142**, 333–353 (1969).

[54] H. Bondi, Bouncing Spheres in General Relativity, *Nature* **215**, 838–839 (1967).

[55] M. Bouhmadi–López, A. Errahmani, P. Martin–Moruno, T. Ouali, and Y. Tavakoli, The little sibling of the big rip singularity, *Int. J. Mod. Phys. D* **24**, 1550078 (2015).

[56] M. Bouhmadi–López, P. F. Gonzalez–Diaz, and P. Martin–Moruno, Worse than a big rip?, *Phys. Lett. B* **659**, 1–5 (2008).

[57] A. J. Bowen, Simple models of near–shore sedimentation, beach profiles and longshore bars, in *The Coastline of Canada: Littoral Processes and Shore Morphology*, edited by S. B. McCann, Geological Survey of Canada, Ottawa (1980), pp. 1–11.

[58] J. Braden, M. C. Johnson, H. V. Peiris, A. Pontzen, and S. Weinfurtner, Nonlinear dynamics of the cold atom analog false vacuum, *J. High Energy Phys.* **10**, 174 (2019).

[59] R. H. Brandenberger, String Gas Cosmology, in *String Cosmology*, edited by J. Erdmenger, Wiley, New York (2009), p. 193–230.

[60] C. Brans and R. H. Dicke, Mach's principle and a relativistic theory of gravitation, *Phys. Rev.* **124**, 925 (1961).

[61] F. Brauer and J. A. Noel, *Introduction to Differential Equations with Applications*, Harper & Row, New York (1986).

[62] C. E. Brennen, *Fundamentals of Multiphase Flows*, Cambridge University Press, Cambridge, UK (2005).

[63] S. Brown and S. Salter, Analogies in science and science teaching, *Adv. Physiol. Edu.* **34**, 167–169 (2010).

[64] W. Brutsaert and J. L. Nieber, Regionalized drought flow hydrographs from a mature glaciated plateau, *Water Res. Research* **13**, 637–643 (1977).

[65] P. Bruun, Coastal Erosion and the Development of Beach Profiles, U.S. Beach Erosion Board Technical Memo 44, p. 79 (1954).

[66] H. A. Buchdahl, Static solutions of the Brans–Dicke equations, *Int. J. Theor. Phys.* **6**, 407 (1972).

[67] E. Bueler, Construction of steady state solutions for isothermal shallow ice sheets, Fairbanks, AK, University of Alaska Fairbanks, Department of Mathematics and Statistics, UAF DMS Tech. Rep. 03–02 (2003).

[68] E. Bueler, C. S. Lingle, J. A. Kallen–Brown, D. N. Covey, and L. N. Bowman, Exact solutions and verification of numerical models for isothermal ice sheets, *J. Glaciol.* **51**, 291–306 (2005).

[69] D. M. Burns and S. G. G. McDonald, *Physics for Biology and Premedical Students*, Addison–Wesley, London (1970).

[70] G. Calcagni, Slow–roll parameters in braneworld cosmologies, *Phys. Rev. D* **69**, 103508 (2004).

[71] K. Caldeira, The maximum entropy principle: a critical discussion, *Clim. Change* **85**, 267–269 (2007).

[72] R. R. Caldwell, A phantom menace? Cosmological consequences of a dark energy component with super–negative equation of state, *Phys. Lett. B* **545**, 23–29 (2002).

[73] R. R. Caldwell, M. Kamionkowski, and N. N. Weinberg, Phantom energy and cosmic doomsday, *Phys. Rev. Lett.* **91**, 071301 (2003).

[74] J. F. Campbell, *Frost and Fire*, J. B. Lippincott, Philadelphia (1865).

[75] S. Capozziello and R. de Ritis, Relation between the potential and nonminimal coupling in inflationary cosmology, *Phys. Lett. A* **177**, 1 (1993).

[76] S. Capozziello and R. de Ritis, Nöther's symmetries and exact solutions in flat non–minimally coupled cosmological models, *Class. Quantum Grav.* **11**, 107 (1994).

[77] S. Capozziello and V. Faraoni, *Beyond Einstein Gravity: A Survey of Gravitational Theories for Cosmology and Astrophysics*, Springer, New York (2010).

[78] S. Capozziello, V. F. Cardone, E. Elizalde, S. Nojiri, and S. D. Odintsov, Observational constraints on dark energy with generalized equations of state, *Phys. Rev. D* **73**, 043512 (2006).

[79] S. Capozziello, V. F. Cardone, E. Elizalde, S. Nojiri, and S. D. Odintsov, Observational constraints on dark energy with generalized equations of state, *Phys. Rev. D* **73**, 043512 (2006).

[80] J. F. Cariñena, M. F. Rañada, and M. Santander, Lagrangian Formalism for nonlinear second–order Riccati Systems: one–dimensional integrability and two–dimensional superintegrability, *J. Math. Phys.* **46**, 062703 (2005).

[81] C. D. Carone, T. V. B. Claringbold, and D. Vaman, Composite graviton self–interactions in a model of emergent gravity, *Phys. Rev. D* **98**, 024041 (2018).

[82] S. M. Carroll, *Spacetime and Geometry: An Introduction to General Relativity*, Addison–Wesley, San Francisco (2004).

[83] A. H. Carter, *Classical and Statistical Thermodynamics*, Prentice Hall, Upper Saddle River, NJ (2001).

[84] S.–Y. Chä and U. R. Fischer, Probing the Scale Invariance of the Inflationary Power Spectrum in Expanding Quasi–Two–Dimensional Dipolar Condensates, *Phys. Rev. Lett.* **118** 130404 (2017); *Erratum, Phys. Rev. Lett.* **118** 179901(E) (2017).

[85] M. P. Chebyshev, *L'intégration des différentielles irrationnelles*, *J. Math. Pures Appl.* **18**, 87 (1853).

[86] S. Chen, G. W. Gibbons, and Y. Yang, Explicit integration of Friedmann's equation with nonlinear equations of state, *J. Cosmol. Astropart. Phys.* **2015**, 020 (2015).

[87] S. Chen, G. W. Gibbons, and Y. Yang, Friedmann–Lemaitre cosmologies via roulettes and other analytic methods, *J. Cosmol. Astropart. Phys.* **10**, 056 (2015).

[88] L. P. Chimento, Symmetry and inflation, *Phys. Rev. D* **65**, 063517 (2002).

[89] J. L. Cieśliński and T. Nikiciuk, A direct approach to the construction of standard and non–standard Lagrangians for dissipative–like dynamical systems with variable coefficients, *J. Phys. A: Math. Theor.* **43**, 175205 (2010).

[90] C. Clanet and D. Quéré, Onset of menisci, *J. Fluid Mech.* **460**, 131–149 (2002).

[91] T. Concannon and G. Giordano, Gravity Tunnel Drag, arXiv:1606.01852 [physics.pop-ph].

[92] P. W. Cooper, Through the Earth in forty minutes, *Am. J. Phys.* **34**, 68–70 (1966).

[93] J. Coté, V. Faraoni, and A. Giusti, Revisiting the conformal invariance of Maxwell's equations in curved spacetime, *Gen. Relativ. Gravit.* **51**, 117 (2019).

[94] C. Criado and N. Alamo, Solving the brachistochrone and other variational problems with soap films, *Am. J. Phys.* **78**, 1400–1405 (2010).

[95] T. Christodoulakis, N. Dimakis, and P. A. Terzis, Lie point and variational symmetries in minisuperspace Einstein gravity, *J. Phys. A* **47**, 095202 (2014).

[96] T. Christodoulakis, A. Karagiorgos, and A. Zampeli, Symmetries in classical and quantum treatment of Einstein's cosmological equations and mini-superspace actions, *Symmetry* **10**, 70 (2018).

[97] K. M. Cuffey and W. S. B. Paterson, *The Physics of Glaciers*, Elsevier, Amsterdam (2010).

[98] M. P. Dabrowski and T. Denkiewicz, Barotropic index w–singularities in cosmology, *Phys. Rev. D* **79**, 063521 (2009).

[99] M. P. Dabrowski, T. Denkiewicz, and M. A. Hendry, How far is it to a sudden future singularity of pressure?, *Phys. Rev. D* **75**, 123524 (2007).

[100] N. Dadhich, Einstein is Newton with space curved, *Current Science* **109** (25), 260–264 (2015).

[101] G. Darmois, *Memorial des Sciences Mathematiques* XXV, Gauthier-Villars, Paris (1927).

[102] R. G. Dean, Equilibrium beach profiles: U.S. Atlantic and Gulf coasts, Department of Civil Engineering, Ocean Engineering Report No. 12, University of Delaware, Newark, DE (1977).

[103] R. G. Dean, Equilibrium beach profiles: Characteristics and applications, *J. Coastal Res.* **1**, 53 (1991).

[104] W. Dean Pesnella, Flying through polytropes, *Am. J. Phys.* **84**, 192 (2016).

[105] W. Dean Pesnella, The flight of Newton's cannonball, *Am. J. Phys.* **86**, 338 (2018).

[106] M. Demianski, R. de Ritis, G. Platania, C. Rubano, P. Scudellaro, and C. Stornaiolo, Scalar field, nonminimal coupling, and cosmology, *Phys. Rev. D* **44**, 3136 (1991).

[107] M. Destrade, G. Gaeta, and G. Saccomandi, Weierstrass' criterion and compact solitary waves, *Phys. Rev. E* **75**, 047601 (2007).

[108] R. C. Dewar, Maximum entropy production and the fluctuation theorem, *J. Phys. A* **38**, L371–L381 (2005).

[109] M. A. De Andrade and L. G. Ferreira Filho, A train that moves using the force of gravity, *Rev. Bras. Ensino Fís.* **40**, 3 (2018).

[110] P.–G. de Gennes, F. Brochard–Wyert, and D. Quéré, *Capillarity and Wetting Phenomena*, Springer, New York (2010).

[111] N. Dimakis, T. Christodoulakis, and T. A. Terzis, FLRW metric cosmology with a perfect fluid by generating integrals of motion, *J. Geom. Phys.* **77**, 97 (2012).

[112] T. J. Dolan and R. G. Dean, Multiple longshore sand bars in the upper Chesapeake Bay, *Estuarine, Coastal and Shelf Science* **21**, 727–743 (1985).

[113] J. Douglas, Solution of the inverse problem of the calculus of variations, *Trans. Am. Math. Soc.* **50**, 71–128 (1941).

[114] M. Dragoni, Gravity in Earth's Interior, *Phys. Teacher* **58**, 97 (2020).

[115] M. Dragoni, I. Borsari, and A. Tallarico, A model for the shape of lava flow fronts, *J. Geophys. Res.* **110**, B09203 (2005).

[116] P. Dulong and A. Petit, Researches on the Measure of Temperatures, and on the Laws of the Communication of Heat, *Ann. Chim. Phys.* **7**, 222–264 (1817).

[117] R. Durrer, *The Cosmic Microwave Background*, Cambridge University Press, Cambridge, UK (2008).

[118] S. Dussault and V. Faraoni, A new symmetry of the spatially flat Einstein–Friedmann equations, *Eur. Phys. J. C* **80**, 1002 (2020).

[119] S. Dussault, V. Faraoni, and A. Giusti, Analogies between Logistic Equation and Relativistic Cosmology, *Symmetry* **13**, 704 (2021).

[120] R. d'Inverno, *Introducing Einstein's General Relativity*, Clarendon Press, Oxford, UK (1998).

[121] S. Eckel, A. Kumar, T. Jacobson, I. B. Spielman, and G. K. Campbell, A Rapidly Expanding Bose–Einstein Condensate: An Expanding Universe in the Lab, *Phys. Rev. X* **8**, 021021 (2018).

[122] A. S. Eddington, On the instability of Einstein's spherical world, *Mon. Not. Roy. Astr. Soc.* **90**, 668–678 (1930).

[123] A. Einstein, *Sitzungsber Akad. Wiss. (Berlin)*, p. 142 (1917). Translated in H. A. Lorentz *et al.*, *Principle of Relativity*, Dover, New York (1952), p. 175.

[124] V. B. Eltsov, T. W. Kibble, M. Krusius, V. M. Ruutu, and G. E. Volovik, Composite defect extends analogy between cosmology and ^{3}He, *Phys. Rev. Lett.* **85**, 4739–4742 (2000).

[125] L. P. Eisenhart, *Riemannian Geometry*, Princeton University Press, Princeton (1949).

[126] G. F. R. Ellis, General Relativity and Cosmology, in *Proceedings of the International School of Physics E. Fermi, Course XLVII*, Varenna, Italy, 1969, edited by R. K. Sachs, Academic Press, New York (1971), pp. 104–182.

[127] G. F. R. Ellis and R. Maartens, The emergent universe: Inflationary cosmology with no singularity, *Class. Quantum Grav.* **21**, 223–232 (2004).

[128] S. H. Emerman and D. L. Turcotte, A fluid model for the shape of accretionary wedges, *Earth Planet. Sci. Lett.* **63**, 379–384 (1983).

[129] J. Erlich, Stochastic Emergent Quantum Gravity, *Class. Quantum Grav.* **35**, 245005 (2018).

[130] C. Eling, R. Guedens, and T. Jacobson, Non–equilibrium thermodynamics of spacetime, *Phys. Rev. Lett.* **96**, 121301 (2006).

[131] L. Euler, Geometria dalya ipotrevleniya v academycheskoi gymnazii, *Novi Comm. Acad. Sci. Petrop.* **11**, 144 (1765).

[132] V. Faraoni, Solving for the dynamics of the universe, *Am. J. Phys.* **67**, 732 (1999).

[133] V. Faraoni, *Cosmology in Scalar–Tensor Gravity*, Kluwer Academic, Dordrecht (2004).

[134] V. Faraoni, Cosmological apparent and trapping horizons, *Phys. Rev. D* **84**, 024003 (2011).

[135] V. Faraoni, A symmetry of the spatially flat Friedmann equations with barotropic fluids, *Phys. Lett. B* **703**, 228 (2011).

[136] V. Faraoni, *Special Relativity*, Springer, New York (2013).

[137] V. Faraoni, Is the Hawking Quasilocal Energy "Newtonian"?, *Symmetry* **7**, 2038–2046 (2015).

[138] V. Faraoni, *Cosmological and Black Hole Apparent Horizons*, Springer, New York (2015).

[139] V. Faraoni, Volume/area scaling of glaciers and ice caps and their longitudinal profiles, *J. Glaciol.* **62**, 928–932 (2016).

[140] V. Faraoni, Solving the Vialov equation of glaciology in terms of elementary functions, *Math. Geosciences* **49**, 1057–1067 (2017).

[141] V. Faraoni, *Alpine Physics: Science In The Mountain Environment*, World Scientific, Singapore (2019).

[142] V. Faraoni, Analogy between equilibrium beach profiles and closed universes, *Phys. Rev. Res.* **1**, 033002 (2019).

[143] V. Faraoni, Modelling the shapes of glaciers: An introduction, *Eur. J. Phys.* **40**, 025802 (2019).

[144] V. Faraoni, Natural phenomena described by the same equation, *Eur. J. Phys.* **41**, 054002 (2020).

[145] V. Faraoni, Cosmic analogues of classic variational problems, *Universe* **6**, 71 (2020).

[146] V. Faraoni, Lagrangian formulation of Omori's law and analogy with the cosmic Big Rip, *Eur. Phys. J. C* **80**, 445 (2020).

[147] V. Faraoni, Analogy between freezing lakes and the cosmic radiation era, *Phys. Rev. Res.* **2**, 013187 (2020).

[148] V. Faraoni, Maximizing friction in the erosion of glacial valleys, *J. Glaciol.* **66**, 876–879 (2020).

[149] V. Faraoni, Lagrangian formulation, a general relativity analogue, and a symmetry of the Vialov equation of glaciology, *Eur. Phys. J. Plus* **135**, 887 (2020).

[150] V. Faraoni, On the extremization of wave energy dissipation rates in equilibrium beach profiles, *J. Oceanography* **76**, 459–463 (2020).

[151] V. Faraoni, A symmetry of the Einstein–Friedmann equations for spatially flat, perfect fluid, universes, *Symmetry* **12**, 147 (2020).

[152] V. Faraoni and F. Atieh, Turning a Newtonian analogy for FLRW cosmology into a relativistic problem, *Phys. Rev. D* **102**, 044020 (2020).

[153] V. Faraoni and F. Atieh, Generalized Fibonacci Numbers, Cosmological Analogies, and an Invariant, *Symmetry* **13**, 200 (2021).

[154] V. Faraoni and A. M. Cardini, Analogues of glacial valley profiles in particle mechanics and in cosmology, *FACETS* **2**, 286–300 (2017).

[155] V. Faraoni and M. W. Vokey, The thickness of glaciers, *Eur. J. Phys.* **36**, 055031 (2015).

[156] V. Faraoni, F. Atieh, and S. Dussault, Cosmological analogies, Lagrangians, and symmetries for convective–radiative heat transfer, *Eur. Phys. J. C* **80**, 706 (2020).

[157] V. Faraoni, S. Jose, and S. Dussault, Multi–fluid cosmology in Einstein gravity: analytical solutions, *Gen. Relativ. Gravit.* **53**, 109 (2021).

[158] V. Faraoni, S. Jose, and A. Leblanc, The curious case of the Buchdahl–Land–Sultana–Wyman–Ibañez–Sanz spacetime, *Phys. Rev. D* **105**, 024030 (2022).

[159] P. O. Fedichev and U. R. Fischer, "Cosmological" quasiparticle production in harmonically trapped superfluid gases, *Phys. Rev. A* **69**, 033602 (2004).

[160] P. O. Fedichev and U. R. Fischer, Gibbons–Hawking Effect in the Sonic de Sitter Space–Time of an Expanding Bose–Einstein–Condensed Gas, *Phys. Rev. Lett.* **91**, 240407 (2007); *Erratum, Phys. Rev. Lett.* **92**, 049901(E) (2007).

[161] M. R. Feldman, J. D. Anderson, G. Schubert, V. Trimble, S. M. Kopeikin, and C. Lämmerzahl, Deep space experiment to measure G, *Class. Quantum Grav.* **33**, 125013 (2016).

[162] J. E. Felten and R. Isaacman, Scale factors $R(t)$ and critical values of the cosmological constant Λ in Friedmann universes, *Rev. Mod. Phys.* **58**, 689 (1986).

[163] L. Fernandez–Jambrina, Hidden past of dark energy cosmological models, *Phys. Lett. B* **656**, 9–14 (2007).

[164] D. Figueiredo, F. Moraes, S. Fumeron, and B. Berche, Cosmology in the laboratory: an analogy between hyperbolic metamaterials and the Milne universe, *Phys. Rev. D* **96**, 105012 (2017).

[165] I. Z. Fisher, Scalar mesostatic field with regard for gravitational effects, *Zh. Eksp. Teor. Fiz.* **18**, 636 (1948).

[166] U. R. Fischer and R. Schützhold, Quantum simulation of cosmic inflation in two–component Bose–Einstein condensates, *Phys. Rev. A* **70**, 063615 (2004).

[167] U. R. Fischer and M. Visser, Riemannian geometry of irrotational vortex acoustics, *Phys. Rev. Lett.* **88**, 110201 (2002).

[168] I. V. Fomin, S. V. Chervon, and S. D. Maharaj, A new look at the Schrödinger equation in exact scalar field cosmology, *Int. J. Geom. Meth. Mod. Phys.* **16**, 1950022 (2018).

[169] A. C. Fowler and D. A. Larson, On the flow of polythermal glaciers. II. Surface wave analysis, *Proc. Roy. Soc. Lond. A* **370**, 155–171 (1980).

[170] P. H. Frampton, K. J. Ludwick, and R. J. Scherrer, The little rip, *Phys. Rev. D* **84**, 063003 (2011).

[171] J.–P. Franc and J.–M. Michel, *Fundamentals of Cavitation*, Kluwer Academic, Dordrecht (2004).

[172] J. Franklin, *Classical Field Theory*, Cambridge University Press, Cambridge, UK (2017).

[173] A. Friedmann, Über die krümmung des raumes, *Z. Phys.* **10**, 377–386 (1922). Translated in Ref. [39] and in: On the curvature of space, *Gen. Relativ. Gravit.* **31**, 1991–2000 (1999).

[174] Y. Fujii and K. Maeda, *The Scalar-Tensor Theory of Gravitation*, Cambridge University Press, Cambridge, UK (2003).

[175] L. J. Garay, J. R. Anglin, J. I. Cirac, and P. Zoller, Sonic Analog of Gravitational Black Holes in Bose–Einstein Condensates, *Phys. Rev. Lett.* **85**, 4643–4647 (2000).

[176] J. Gardiner, Fibonacci, quasicrystals and the beauty of flowers, *Plant Signaling & Behavior* **7**, 1721 (2012).

[177] R. Garra, private communication (2020).

[178] W. A. Gaskell, A laboratory study of the inductive theory of thunderstorm electrification, *Quart. J. Roy. Meteorol. Soc.* **107**, 955–966 (1981).

[179] G. Gionti, and A. Paliathanasis, Duality transformation and conformal equivalent scalar–tensor theories, *Mod. Phys. Lett. A* **33**, 1850093 (2018).

[180] A. Gjerløv and D. Dean Pesnella, Orbits through polytropes, *Am. J. Phys.* **87**, 452 (2019).

[181] E. N. Glass, Shear-free gravitational collapse, *J. Math. Phys.* **20**, 1508 (1979).

[182] J. W. Glen, The creep of polycrystalline ice, *Proc. Roy. Soc. Lond. A* **228**, 519–538 (1955).

[183] H. Goldstein, *Classical Mechanics*, Addison–Wesley, Reading, Massachusetts (1980).

[184] R. Goody, Maximum entropy production in climate theory, *J. Atmos. Sci.* **64**, 2735–2739 (2007).

[185] V. Gorini, A. Kamenshchik, U. Moschella, and V. Pasquier, Tachyons, scalar fields and cosmology, *Phys. Rev. D* **69**, 123512 (2004).

[186] R. Greve and H. Blatter, *Dynamics of Ice Sheets and Glaciers*, Springer, New York (2009).

[187] R. W. Griffiths, The dynamics of lava flows, *Annu. Rev. Fluid Mech.* **32**, 477–518 (2000).

[188] G. Grinstein and R. Linsker, Comments on a derivation and application of the maximum entropy production principle, *J. Phys. A* **40**, 9717–9720 (2007).

[189] E. I. Guendelman and A. Rabinowitz, Gravitational field of a hedgehog and the evolution of vacuum bubbles, *Phys. Rev. D* **44**, 3152 (1991).

[190] A. V. Guglielmi, Interpretation of the Omori law, *Izv. Phys. Solid Earth* **52**, 785–786 (2016).

[191] A. V. Guglielmi, Omori's law: A note on the history of geophysics, *Physics Uspekhi* **60**, 319–324 (2017).

[192] A. V. Guglielmi and A. D. Zavyalov, The 150th Anniversary of Fusakichi Omori, arXiv:1803.08555 [physics.geo-ph].

[193] A. H. Guth, Inflationary universe: A possible solution to the horizon and flatness problems, *Phys. Rev. D* **23**, 347 (1981).

[194] H. Haken, *Synergetics–An Introduction*, Springer, Berlin (1978).

[195] J. S. Han, H. S. Kim, and J. Neggers, On Fibonacci functions with Fibonacci numbers, *Advances in Difference Equations* **2012**, 126 (2012).

[196] W. Hao, Z. Wu, H. Xia, Y. Bo, G. Xin, and C. Ling, Numerical study of influence of deep coring parameters on temperature of in–situ core, *Thermal Science* **23**, 1441–1447 (2019).

[197] J. Harbor, A discussion of Hirano and Aniya's (1988, 1989) explanation of glacial–valley cross profile development, *Earth Surf. Proc. Landforms* **15**, 369–377 (1990).

[198] J. M. Harbor, Development of glacial–valley cross sections under conditions of spatially variable resistance to erosion, *Geomorphol.* **14**, 99–107 (1995).

[199] T. Harko, F. S. N. Lobo, and M. K. Mak, A Riccati equation based approach to isotropic scalar field cosmologies, *Int. J. Mod. Phys. D* **23**, 1450063 (2014).

[200] A. J. L. Harris, J. Bailey, S. Calvari, and J. Dehn, Heat loss measured at a lava channel and its implications for down–channel cooling and rheology, *Geol. Soc. Am. Spec. Pap.* **396**, 125–146 (2005).

[201] S. Hawking, Gravitational Radiation in an Expanding Universe, *J. Math. Phys.* **9**, 598–604 (1968).

[202] S. W. Hawking and G. F. R. Ellis, *The Large Scale Structure of Space–Time*, Cambridge University Press, Cambridge, UK (1973).

[203] R. M. Hawkins and J. E. Lidsey, The Ermakov–Pinney equation in scalar field cosmologies, *Phys. Rev. D* **66**, 023523 (2002).

[204] S. A. Hayward, Quasilocal gravitational energy, *Phys. Rev. D* **49**, 831–839 (1994).

[205] S. A. Hayward, Gravitational energy in spherical symmetry, *Phys. Rev. D* **53**, 1938–1949 (1996).

[206] W. C. Hernandez and C. W. Misner, Observer time as a coordinate in relativistic spherical hydrodynamics, *Astrophys. J.* **143**, 452 (1966).

[207] E. R. Harrison, Cosmology without General Relativity, *Ann. Phys. (NY)* **35**, 437–446 (1965).

[208] D. Heggie and P. Hut, *The Gravitational Million–Body Problem*, Cambridge University Press, Cambridge, UK (2003), pp. 143–149.

[209] H. von Helmholtz, Ueber die physikalische Bedeutung des Princips der kleinsten Wirkung, *Journal für die reine und angewandte Mathematik* **100**, 18 (1887).

[210] E. Hille, *Lectures on Ordinary Differential Equations*, Addison–Wesley, Reading, Massachusetts (1969).

[211] S. Hirano, Investigation of aftershocks of the great Kanto earthquake at Kumagawa, *Kishosushi*, *Ser.* 2, **2**, 77 (1924) (in Japanese).

[212] M. Hirano and M. Aniya, A rational explanation of cross–profile morphology for glacial valleys and of glacial valley development, *Earth Surf. Proc. Landforms* **13**, 707–716 (1988).

[213] M. Hirano and M. Aniya, A reply to 'A discussion of Hirano and Aniya's (1988, 1989) explanation of glacial–valley cross profile development' by Jonathan M. Harbor, *Earth Surf. Proc. Landforms* **15**, 379–381 (1990).

[214] M. Hirano and M. Aniya, Response to Morgan's comment, *Earth Surf. Proc. Landforms* **30**, 515 (2005).

[215] V. E. Hoggatt, Jr. and M. Bicknell–Johnson, Fibonacci Convolution Sequences, *Fib. Quart.* **15**, 117 (1977).

[216] R. L. B. Hooke, *Principles of Glacier Mechanics*, 2nd edition, Cambridge University Press, Cambridge, UK (2005).

[217] S. Hossenfelder, Covariant version of Verlinde's emergent gravity, *Phys. Rev. D* **95**, 124018 (2017).

[218] T. W. Hsu, I. F. Tseng, and C. P. Lee, A new shape function for bar–type beach profiles, *J. Coastal Res.* **22**, 728–736 (2006).

[219] J.–D. Huang, D. W. T. Jackson, and J. A. G. Cooper, Piecewise Polynomial Expression of Beach Profiles, *J. Coastal Res.* **265**, 851–859 (2010).

[220] V. E. Hubeny, The AdS/CFT Correspondence, *Class. Quantum Grav.* **32**, 124010 (2015).

[221] K. Hutter, *Theoretical Glaciology*, Reidel, Dordrecht (1983).

[222] E. L. Ince, *Ordinary Differential Equations*, Dover, New York (1944).

[223] D. L. Inman, M. H. S. Elwany, and S. A. Jenkins, Shore rise and bar–berm profiles on ocean beaches, *J. Geophys. Res.* **98 (C10)**, 18181–18199 (1993).

[224] S. Isermann, Analytical solution of gravity tunnels through an inhomogeneous Earth, *Am. J. Phys.* **87**, 10 (2019).

[225] S. Isermann, Free fall through the rotating and inhomogeneous Earth, *Am. J. Phys.* **87**, 646 (2019).

[226] W. Israel, Singular hypersurfaces and thin shells in general relativity, *Nuovo Cimento B* **44**, 1 (1966); *Erratum* **48**, 463(E) (1967).

[227] T. Jacobson, Thermodynamics of Spacetime: The Einstein Equation of State, *Phys. Rev. Lett.* **75**, 1260–1263 (1995).

[228] T. Jacobson, When is $g_{tt} \, g_{rr} = -1$?, *Class. Quantum Grav.* **24**, 5717 (2007).

[229] T. A. Jacobson and G. E. Volovik, Effective spacetime and Hawking radiation from moving domain wall in thin film of ^{3}He-A, *J. Exp. Theor. Phys. Lett.* **68**, 874–880 (1998).

[230] P. Jacquod, P. G. Silvestrov, and C. W. J. Beenakker, Golden rule decay versus Lyapunov decay of the quantum Loschmidt echo, *Phys. Rev. E* **64**, 055203(R) (2001).

[231] A. J. Jaeger, A. R. Taylor, and J. Wiley, When, and for whom, analogies help: The role of spatial skills and interleaved presentation, *J. Educ. Psychol.* **108**, 1121–1139 (2016).

[232] A. I. Janis, E. T. Newman, and J. Winicour, Reality of the Schwarzschild Singularity, *Phys. Rev. Lett.* **20**, 878 (1968).

[233] R. T. Jantzen and C. Uggla, The Structure of the Generalized Friedmann Problem, *Gen. Relativ. Gravit.* **24**, 59–85 (1992).

[234] H. Jeffreys, Aftershocks and periodicity in earthquakes, *Gerlands Beitr. Geophys.* **86**, 111 (1938).

[235] M. O. Jeffries, K. Morris, and C. R. Duguay, Lake ice growth and decay in central Alaska, USA: observations and computer simulations compared, *Ann. Glaciol.* **40**, 195–199 (2005).

[236] S. A. Jenkins and D. L. Inman, Thermodynamic solutions for equilibrium beach profiles, *J. Geophys. Res.: Oceans* **111**, C02003 (2006).

[237] J. N. Johnson, Kinematic Waves and Group Velocity: Applications to Natural and Man–Made Environment, *Am. J. Phys.* **42**, 681 (1974).

[238] D. E. Kelly, Convection in ice–covered lakes: effects on algal suspension, *J. Plankton Res.* **19/12**, 1859–1880 (1997).

[239] P. G. Kirmser, An example of the need for adequate references, *Am. J. Phys.* **4**, 701 (1966).

[240] V. V. Kiselev, Quintessence and black holes, *Class. Quantum Grav.* **20**, 1187 (2003).

[241] A. R. Klotz, The gravity tunnel in a non–uniform Earth, *Am. J. Phys.* **83**, 231 (2015).

[242] A. R. Klotz, A Guided Tour of Planetary Interiors, arXiv:1505.05894 [physics.pop-ph].

[243] G. Kofinas, R. Maartens, and E. Papantonopoulos, Brane cosmology with curvature corrections, *J. High Energy Phys.* **2003**, 066 (2003).

[244] E. W. Kolb and M. S. Turner, *The Early Universe*; Addison–Wesley, Redwood City, CA, USA (1990).

[245] E. B. Kolomeisky, Natural analog to cosmology in basic condensed matter physics, *Phys. Rev. B* **100**, 140301 (2019).

[246] P. D. Komar and W. G. McDougal, The analysis of exponential beach profiles, *J. Coastal Res.* **10**, 59–69 (1994).

[247] H. W. Kroto, J. R. Heath, S. C. O'Brien, R. F. Curl, and R. E. Smalley, C 60: Buckminsterfullerene, *Nature* **318**, 162–163 (1985).

[248] G. Lamé and B. T. E. Clapyeron, Memoire sur la solidification par refroidissment d'un globe solid, *Ann. Chim. Phys.* **47**, 250–260 (1831).

[249] C. Lanczos, Flächenhafte Verteilung der Materie in der Einsteinschen Gravitationstheorie, *Ann. Phys. (Leipzig)* **379**, 518 (1924).

[250] L. D. Landau and E. M. Lifschitz, *The Classical Theory of Fields*, Pergamon Press, Oxford, UK (1989).

[251] M. Larson and N. C. Kraus, Temporal and spatial scales of beach profile change, Duck, North Carolina, *Marine Geology* **117**, 75–94 (1994).

[252] M. Larson, N. C. Kraus, and A. R. Wise, Equilibrium beach profiles under breaking and non–breaking waves, *Coastal Engineering* **36**, 59–85 (1999).

[253] L. J. Laslett, Trajectory for minimum transit time through the earth, *Am. J. Phys.* **34**, 702–703 (1966).

[254] J. Launiainen and B. Cheng, Modelling of ice thermodynamics in natural water bodies, *Cold Reg. Sci. Technol.* **27**, 153–178 (1998).

[255] G. Lemaître, Un Univers homogéne de masse constante et de rayon croissant rendant compte do la vitesse radiale des nébuleuses extra–galactiques, *Ann. Soc. Sci. Bruxelles* **A 47**, 49–59 (1927). Translated as: A homogeneous universe of constant mass and increasing radius accounting for the radial velocity of extra–galactic nebulae, *Mon. Not. Roy. Astr. Soc.* **91**, 483–490 (1931).

[256] N. Lemarchand and J.–R. Grasso, Interactions between earthquakes and volcano activity, *Geophys. Res. Lett.* **34**, L24303 (2007).

[257] M. Leppäranta, Modelling the formation and decay of lake ice, in *The Impact of Climate Change on European Lakes* (Aquatic Ecology Series 4) edited by D. G. George, Springer, Berlin (2010).

[258] T. Levi–Civita, Sur la régularisation du problème des trois corps, *Acta Math.* **42**, 99–144 (1920).

[259] S. Liberati, Analogue gravity models of emergent gravity: lessons and pitfalls, *J. Phys. Conf. Ser.* **880**, 012009 (2017).

[260] A. R. Liddle, *An Introduction to Modern Cosmology*, Wiley, New York (2015).

[261] A. R. Liddle and D. H. Lyth, *Cosmological Inflation and Large–Scale Structure*, Cambridge University Press, Cambridge, UK (2000).

[262] M. J. Lighthill and G. B. Whitham, On kinematic waves. I. Flood movement in long rivers, *Proc. Roy. Soc. A* **229**, 281–316 (1955).

[263] M. J. Lighthill and G. B. Whitham, On kinematic waves. II. A theory of traffic flow on long crowded roads, *Proc. Roy. Soc. A* **229**, 317–345 (1955).

[264] A. Linde, *Particle Physics and Inflationary Cosmology*, Harwood Academic, Chur, Switzerland (1990).

[265] G. E. Liston and D. K. Hall, An energy–balance model of lake–ice evolution, *J. Glaciol.* **41/138**, 373–382 (1995).

[266] J. Lopuszanski, *The Inverse Variational Problem in Classical Mechanics*, World Scientific, Singapore (1999).

[267] J.–P. Luminet, J. Weeks, A. Riazuelo, R. Lehoucq, and J.–P. Uzan, Dodecahedral space topology as an explanation for weak wide–angle temperature correlations in the cosmic microwave background, *Nature* **425**, 593 (2003).

[268] S. Maldonado, Do beach profiles under non–breaking waves minimize energy dissipation?, *J. Geophys. Res.: Oceans* **125**, e2019JC015876 (2020).

[269] S. Maldonado and M. Uchasara, On the thermodynamics–based equilibrium beach profile derived by Jenkins and Inman (2006), arXiv:1908.07825 [physics.geo-ph].

[270] R. L. Mallett, Comments on "Through the Earth in forty minutes", *Am. J. Phys.* **34**, 702 (1966).

[271] S. C. Mancas and H. C. Rosu, Evolution of spherical cavitation bubbles: parametric and closed–form solutions, *Phys. Fluids* **28**, 022009 (2016).

[272] R. Mansouri, On the non–existence of time–dependent fluid spheres in general relativity obeying an equation of state, *Ann. Inst. Henri Poincaré*, **A27**, 175 (1977).

[273] E. A. Marchisotto and G.–A. Zakeri, An invitation to integration in finite terms, *College Math. J.* **25**, 295 (1994).

[274] B. Mashhoon and M. H. Partovi, On the gravitational motion of a fluid obeying an equation of state, *Ann. Phys. (NY)* **130**, 99–138 (1980).

[275] B. J. Mason, *The Physics of Clouds*, 2nd edition, Oxford University Press, Oxford, UK (1971).

[276] B. J. Mason, The generation of electric charges and fields in thunderstorms, *Proc. Roy. Soc. A* **415**, 303–315 (1988).

[277] B. J. Mason and N. Mason, The physics of a thunderstorm, *Eur. J. Phys.* **24**, S99–S110 (2003).

[278] D. M. McClung, Derivation of Voellmy's Maximum Speed and Run–Out Estimates from a Centre–of–Mass Model, *J. Glaciol.* **29**, 350–352 (1983).

[279] W. J. McGee, Glacial Canons, *J. Geology* **2**, 350–364 (1894).

[280] C. P. McKay, G. D. Clow, R. A. Wharton, and S. W. Squyres, Thickness of ice on perennially frozen lakes, *Nature* **313**, 561–562 (1985).

[281] G. C. McVittie, An example of gravitational collapse in general relativity, *Astrophys. J.* **143**, 682 (1966).

[282] Merriam–Webster Dictionary online, https://www.merriam-webster.com

[283] C. W. Misner and D. H. Sharp, Relativistic Equations for Adiabatic, Spherically Symmetric Gravitational Collapse, *Phys. Rev.* **136**, B571 (1964).

[284] B. L. Moiseiwitsch, *Variational Principles*, Interscience, London, UK (1966).

[285] J. A. Montemayor–Aldrete, J. D. Muñoz–Andrade, A. Mendoza–Allende, and A. Montemayor–Varela, Analogy betwen dislocation creep and relativistic cosmology, *Rev. Mexican. Fis.* **51**, 461–475 (2005).

[286] F. Morgan, A note on cross–profile morphology for glacial valleys, *Earth Surf. Proc. Landforms* **30**, 513–514 (2005).

[287] M. S. Morris and K. S. Thorne, Wormholes in spacetime and their use for interstellar travel: A tool for teaching General Relativity, *Am. J. Phys.* **56**, 395–412 (1988).

[288] V. Mukhanov, *Physical Foundations of Cosmology*, Cambridge University Press, Cambridge, UK (2005).

[289] J. R. Muñoz de Nova, K. Golubkov, V. I. Kolobov, and J. Steinhauer, Observation of thermal Hawking radiation and its temperature in an analogue black hole, *Nature* **569**, 688–691 (2019).

[290] Z. E. Musielak, Standard and non–standard Lagrangians for dissipative dynamical systems with variable coefficients, *J. Phys. A: Math. Theor.* **41**, 055205 (2008).

[291] Z. E. Musielak, D. Roy, and L. D. Swift, Method to derive Lagrangian and Hamiltonian for a nonlinear dynamical system with variable coefficients, *Chaos, Solitons & Fractals* **38**, 894–902 (2008).

[292] M. Nelkon and P. Parker, *Advanced Level Physics*, Heinemann, London, UK (1977).

[293] F. Ng, and E. C. King, Kinematic waves in polar firn stratigraphy, *J. Glaciol.* **57**, 1119–1134 (2011).

[294] F. S. L. Ng, I. D. Barr, and C. D. Clark, Using the surface profiles of modern ice masses to inform palaeo–glacier reconstruction, *Quat. Sci. Rev.* **29**, 3240–3255 (2010).

[295] C. Nicolis and G. Nicolis, Stability, complexity and the maximum dissipation conjecture, *Quart. J. Roy. Meteorol. Soc.* **136**, 1161–1169 (2010).

[296] A. B. Nielsen and M. Visser, Production and decay of evolving horizons, *Class. Quantum Grav.* **23**, 4637 (2006).

[297] S. Nojiri and S. D. Odintsov, Final state and thermodynamics of a dark energy universe, *Phys. Rev. D*, **70**, 103522 (2004).

[298] S. Nojiri and S. D. Odintsov, Inhomogeneous equation of state of the universe: phantom era, future singularity, and crossing the phantom barrier, *Phys. Rev. D* **72**, 023003 (2005).

[299] S. Nojiri, S. D. Odintsov, and S. Tsujikawa, Properties of singularities in the (phantom) dark energy universe, *Phys. Rev. D* **71**, 063004 (2005).

[300] M. Novello, M. Visser, and G. Volovik (editors), *Artificial Black Holes*, World Scientific, Singapore (2002).

[301] M. Novakowski and H. C. Rosu, Newton's laws of motion in form of Riccati equation, *Phys. Rev. E* **65**, 047602 (2002).

[302] J. F. Nye, The flow of glaciers and ice–sheets as a problem in plasticity, *Proc. Roy. Soc. Lond. A* **207**, 554–572 (1951).

[303] J. F. Nye, A method of calculating the thicknesses of the ice–sheets, *Nature* **169**, 529–530 (1951).

[304] H. F. Olson, *Dynamical Analogies*, Van Nostrand, Princeton, NJ, USA (1959).

[305] F. J. Omori, On the aftershocks of earthquakes, *J. Coll. Sci. Imperial Univ. Tokyo* **7**, 111–200 (1894).

[306] R. Oppenheimer, *Analogy in Science*, The Centennial Review of Arts & Science, Vol. 2, Michigan State University Press (1958), pp. 351–373.

[307] J. R. Oppenheimer and J. R. Snyder, On continued gravitational contraction, *Phys. Rev.* **56**, 455–459 (1939).

[308] M. Orgill and G. Bodner, What research tells us about using analogies to teach chemistry, *Chem. Educ, Res. Practice* **5**, 15–32 (2004).

[309] D. Oriti, Levels of spacetime emergence in quantum gravity, arXiv:1807.04875 [physics.hist-ph].

[310] E. G. Otvos, Beach ridges—definitions and significance, *Geomorphol.* **32**, 83–108 (2000).

[311] C. T. O'Sullivan, Newton's law of cooling—A critical assessment, *Am. J. Phys.* **58**, 956–960 (1990).

[312] T. Padmanabhan, *Structure Formation In The Universe*, Cambridge University Press, Cambridge, UK (1993).

[313] T. Padmanabhan, Thermodynamical aspects of gravity: new insights, *Rep. Prog. Phys.* **73**, 046901 (2010).

[314] T. Pailas, N. Dimakis, A. Paliathanasis, P. A. Terzis, and T. Christodoulakis, The infinite dimensional symmetry groups of the Friedmann equation, *Phys. Rev. D* **102**, 063524 (2020).

[315] A. Paliathanasis and S. Capozziello, Noether symmetries and duality transformations in cosmology, *Mod. Phys. Lett. A* **31**, 1650183 (2016).

[316] A. Paliathanasis, M. Tsamparlis, and S. Basilakos, Constraints and analytical solutions of $f(R)$ theories of gravity using Noether symmetries, *Phys. Rev. D* **84**, 123514 (2011).

[317] A. Paliathanasis, M. Tsamparlis, S. Basilakos, and J. D. Barrow, Classical and quantum solutions in Brans–Dicke cosmology with a perfect fluid, *Phys. Rev. D* **93**, 043528 (2016).

[318] A. C. Palmer, A kinematic wave model of glacier surges, *J. Glaciol.* **61**, 65–72 (1972).

[319] E. Parker, A relativistic gravity train, *Gen. Relativ. Gravit.* **49**, 106 (2017).

[320] P. Parsons and J. D. Barrow, New inflation from old, *Class. Quantum Grav.* **12**, 1715 (1995).

[321] O. K. Pashaev and J.-H. Lee, Resonance Solitons as Black Holes in Madelung Fluid, *Mod. Phys. Lett. A* **17**, 1601–1619 (2002).

[322] W. S. B. Paterson, Laurentide ice sheet: estimated volume during late Wisconsin, *Rev. Geophys. Space Phys.* **10**, 885–917 (1972).

[323] W. S. B. Paterson, *The Physics of Glaciers*, 3rd edition, Butterworth–Heinemann, Oxford, UK (1994).

[324] S. Patrick, A. Coutant, M. Richartz, and S. Weinfurtner, Black Hole Quasibound States from A Draining Bathtub Vortex Flow, *Phys. Rev. Lett.* **121**, 061101 (2018).

[325] S. Patrick, H. Goodhew, C. Gooding, and S. Weinfurtner, Backreaction in an analogue black hole experiment, *Phys. Rev. Lett.* **126**, 041105 (2021).

[326] J. C. Patterson and P. F. Hamblin, Thermal simulation of a lake with winter ice cover, *Limnol. Oceanogr.* **33/3**, 323–338 (1988).

[327] F. Pattyn and W. Van Huele, Power law or power flaw?, *Earth Surf. Proc. Landforms* **23**, 761–767 (1998).

[328] P. J. E. Peebles, *Principles of Physical Cosmology*, Princeton University Press, Princeton (1993).

[329] R. Perla, T. T. Cheng, and D. M. McClung, A two–parameter model for snow–avalanche motion, *J. Glaciol.* **26**, 197–207 (1980).

[330] M. Pilakouta, N. Kallithrakas–Kontos, and G. Nikolaou, Determining the ^{40}K radioactivity in rocks using X–ray spectrometry, *Eur. J. Phys.* **38**, 055803 (2017).

[331] C. M. A. Pinto, A. Mendes Lopes, and J. A. Tenreiro Machado, A review of power laws in real life phenomena, *Comm. Nonlinear Sci. Num. Simul.* **17**, 3558–3578 (2012).

[332] M. S. Plesset, The dynamics of cavitation bubbles, *J. Appl. Mech.* **16**, 277–282 (1949).

[333] G. Polya, *Mathematics and Plausible Reasoning. Volume 1: Induction and Analogy in Mathematics*, Princeton University Press, Princeton (1954).

[334] M. A. C. Potenza, The daylight sky and Avogadro's number, *Eur. J. Phys.* **36**, 065040 (2015).

[335] A. Prain, S. Fagnocchi, and S. Liberati, Analogue cosmological particle creation: Quantum correlations in expanding Bose–Einstein condensates, *Phys. Rev. D* **82**, 105018 (2010).

[336] A. Prain, C. Maitland, D. Faccio, and F. Marino, Superradiant scattering in fluids of light, *Phys. Rev. D* **100**, 024037 (2019).

[337] G. E. Prince and D. King, The inverse problem in the calculus of variations: nonexistence of Lagrangians, in *Differential Geometric Methods on Mechanics and Field Theory: Volume in Honour of Willy Sarlet*, F. Cantrijn and B. Langerock editors, Academia Press, Gent, Belgium (2007), pp. 131–140.

[338] S. P. Pudasaini and K. Hutter, *Avalanche Dynamics: Dynamics of Rapid Flows of Dense Granular Avalanches*, Springer, Berlin (2007).

[339] S. P. Pudasaini and M. Krautblatter, The Landslide Velocity, *Earth Surf. Dynamics* **10**, 165–189 (2022).

[340] L. Rayleigh, VIII. On the pressure developed in a liquid during the collapse of a spherical cavity, *Phil. Mag.* **34**, 94–98 (1917).

[341] R. Repnik, A. Ranjkesh, V. Simonka, M. Ambrozic, Z. Bradac, and S. Kralj, Symmetry breaking in nematic liquid crystals: analogy with cosmology and magnetism, *J. Phys.: Cond. Matter* **25**, 404201 (2013).

[342] W. Rindler, *Relativity: Special, General and Cosmological*, Oxford University Press, Oxford, UK (2001).

[343] H. C. Rosu, Darboux class of cosmological fluids with time–dependent adiabatic indices, *Mod. Phys. Lett. A* **15**, 979–990 (2000).

[344] H. C. Rosu and V. Ibarra–Junquera, FRW barotropic zero modes: Dynamical systems observability, *Applied Math. Sciences* **1**, 843–852 (2007).

[345] H. C. Rosu and K. V. Khmelnytskaya, Shifted Riccati procedure: Application to conformal barotropic FRW cosmologies, *SIGMA* **7**, 013 (2011).

[346] H. C. Rosu and K. V. Khmelnytskaya, Inhomogeneous barotropic FRW cosmologies in conformal time, *Mod. Phys. Lett. A* **28**, 1340017 (2013).

[347] H. C. Rosu and P. Ojeda–May, Supersymmetry of FRW barotropic cosmologies, *Int. J. Theor. Phys.* **45**, 1152–1157 (2006).

[348] H. C. Rosu, S. C. Mancas, and P. Chen, Barotropic FRW cosmologies with Chiellini damping in comoving time, *Mod. Phys. Lett. A* **30**, 1550100 (2015).

[349] W. W. Rouse Ball, *A Short Account of the History of Mathematics*, Dover, New York (1960).

[350] G. Rousseaux and S. C. Mancas, Visco–elastic cosmology for a sparkling universe?, *Gen. Relativ. Gravit.* **52**, 55 (2020).

[351] E. J. Routh, *A Treatise on Dynamics of a Particle*, Cambridge University Press, Cambridge, UK (1898).

[352] S. K. Roy, P. Baral, R. Koley, and P. Majumdar, Granular Matter as a Gravity Analogue, arXiv:2011.01194 [cond-mat.soft].

[353] B. G. Ruessink and J. H. J. Terwindt, The behavior of nearshore bars on the time scale of years: a conceptual model, *Marine Geology* **163**, 289–302 (2000).

[354] B. Ryden, *Introduction to Cosmology*, Addison Wesley, San Francisco (2003).

[355] A. Saha and B. Talukdar, Inverse variational problem for nonstandard Lagrangians, *Repts. Math. Phys.* **73**, 299–309 (2014).

[356] V. Sahni and Y. Shtanov, Unusual cosmological singularities in brane world models, *Class. Quantum Grav.* **19**, L101–L107 (2002).

[357] D. S. Salopek and J. R. Bond, Nonlinear evolution of long–wavelength metric fluctuations in inflationary models, *Phys. Rev. D* **42**, 3936 (1990).

[358] A. K. Sanyal and B. Modak, Is Noether Symmetric Approach Consistent With Dynamical Equation In Non–minimal Scalar–Tensor Theories?, *Class. Quantum Grav.* **18**, 3767 (2001).

[359] W. Sarlet, The Helmholtz conditions revisited. A new approach to the inverse problem of Lagrangian dynamics, *J. Phys. A: Math. Gen.* **15**, 1503–1517 (1982).

[360] D. Schuch, Nonlinear Riccati equations as a unifying link between linear quantum mechanics and other fields of physics, *J. Physics Conf. Ser.* **504**, 012005 (2014).

[361] D. Schuch, *Quantum Theory From a Nonlinear Perspective*, Fundamental Theories of Physics vol. 191, Springer, New York (2018).

[362] M. Seel, The relativistic gravity train, *Eur. J. Phys.* **39**, 3 (2018).

[363] M. Selmkea, A note on the history of gravity tunnels, *Am. J. Phys.* **86**, 153 (2018).

[364] Y. Shtanov and V. Sahni, New cosmological singularities in braneworld models, *Class. Quantum Grav.* **19**, L101–L107 (2002).

[365] A. Schmid and J.-R. Grasso, Omori law for eruption foreshocks and aftershocks, *J. Geophys. Res.* **117**, B07302 (2012).

[366] L. I. Schiff, On Experimental Tests of the General Theory of Relativity. *Am. J. Phys.* **28**, 340–343 (1960).

[367] R. Schützhold, G. Plunien, and G. Soff, Dielectric Black Hole Analogs, *Phys. Rev. Lett.* **88**, 061101 (2002).

[368] R. Schützhold and W. G. Unruh, Gravity wave analogues of black holes, *Phys. Rev. D* **66**, 044019 (2002).

[369] R. Schützhold and W. G. Unruh, Hawking Radiation in an Electromagnetic Waveguide?, *Phys. Rev. Lett.* **95**, 031301 (2005).

[370] H. Seddik, R. Greve, and S. Sugiyama, Numerical simulation of the evolution of glacial valley cross sections, arXiv:0901.1177 [physics.geo-ph].

[371] L. E. Sigler, *Fibonacci's Liber Abaci: A Translation into Modern English of Leonardo Pisano's Book of Calculation*, Springer, New York (2002).

[372] S. Silva e Costa, An entirely analytical cosmological model, *Mod. Phys. Lett. A* **24**, 531–540 (2009).

[373] A. Simonič, A note on a straight gravity tunnel through a rotating body, *Am. J. Phys.* **88**, 499–502 (2020).

[374] T. R. Slatyer and C. M. Savage, Superradiant scattering from a hydrodynamic vortex, *Class. Quantum Grav.* **22**, 3833–3839 (2005).

[375] J. Smoller and B. Temple, Shock–wave solutions in closed form and the Oppenheimer–Snyder limit in General Reality, *SIAM J. Appl. Math.* **58**, 15–33 (1998).

[376] I. I. Smolyaninov, Linear and nonlinear optics of surface plasmon toy–models of black holes and wormholes, *Phys. Rev. B* **69**, 205417 (2004).

[377] I. Soletta and M. Branca, The frozen lake: a physical model using calculator–based laboratory technology, *Phys. Teacher* **43**, 214–217 (2005).

[378] C. Somigliana, Sulla profondità dei ghiacciai, Nota I, *Rend. Mat. Acc. Lincei Ser. 5* **30-1**, 291–296 (1921).

[379] C. Somigliana, Sulla profondità dei ghiacciai, Nota II, *Rend. Mat. Acc. Lincei Ser. 5* **30-1**, 323–327 (1921).

[380] C. Somigliana, Sulla profondità dei ghiacciai, Nota III, *Rend. Mat. Acc. Lincei Ser. 5* **30-1**, 360–364 (1921).

[381] C. Somigliana, Sulla profondità dei ghiacciai, Nota IV, *Rend. Mat. Acc. Lincei Ser. 5* **30-2**, 3–7 (1921).

[382] S. Sonego and V. Talamini, Qualitative study of perfect–fluid Friedmann–Lemaître–Robertson–Walker models with a cosmological constant, *Am. J. Phys.* **80**, 670–679 (2012).

[383] D. C. Srivastava and S. S. Prasad, Perfect Fluid Spheres in General Relativity, *Gen. Relativ. Gravit.* **15**, 65–77 (1983).

[384] A. A. Starobinsky, A new type of isotropic cosmological models without singularity, *Phys. Lett. B* **91**, 99–102 (1980).

[385] J. Stefan, Ueber die theorie der eisbildung, insbesondere über die Eisbildung im Polarmeere, *Ann. Phys. (Leipzig)* **278/2**, 269–286 (1891).

[386] H. Štefančić, Expansion around the vacuum equation of state: sudden future singularities and asymptotic behavior, *Phys. Rev. D* **71**, 084024 (2005).

[387] H. Stephani, D. Kramer, M. MacCallum, C. Hoenselaers, and E. Hertl, *Exact Solutions of Einstein's Field Equations*, 2nd edition, Cambridge University Press, Cambridge, UK (2003).

[388] S. H. Strogatz, *Nonlinear Dynamics and Chaos: With Applications in Physics, Biology, Chemistry, and Engineering*, Addison–Wesley, Reading, Massachusetts (1994).

[389] J. Sultana, Generating time dependent conformally coupled Einstein–scalar solutions, *Gen. Relativ. Gravit.* **47**, 73 (2015).

[390] K. Sundman, Mémoire sur le problème des trois corps, *Acta Math.* **36**, 105–179 (1912).

[391] S. Surya, Evidence for the continuum in 2d causal set quantum gravity, *Class. Quantum Grav.* **29**, 132001 (2012).

[392] S. Surya, The causal set approach to quantum gravity, *Living Rev. Relativ.* **22**, 5 (2019).

[393] H. Svensson, Is the cross–section of a glacial valley a parabola?, *J. Glaciol.* **3**, 362–363 (1959).

[394] L. B. Szabados, Quasi–local energy–momentum and angular momentum in general relativity, *Living Rev. Relativ.* **12**, 4 (2009).

[395] M. Szydlowski and M. Heller, Equation of state and equation symmetries in cosmology, *Acta Phys. Polon.* **B14**, 571–580 (1983).

[396] M. Szydlowski and O. Hrycyna, Dissipative or conservative cosmology with dark energy?, *Ann. Phys. (NY)* **322**, 2745–2775 (2007).

[397] M. Szydlowski, W. Godlowski, and R. Wojtak, Equation of state for the Universe from similarity symmetries, *Gen. Relativ. Gravit.* **38**, 795–821 (2006).

[398] M. Szydlowski, A. Stachowski, A. Borowiec, and A. Wojnar, Do sewn up singularities falsify the Palatini cosmology?, *Eur. Phys. J. C* **76**, 567 (2016).

[399] R. Taillet, Free falling inside flattened spheroids: Gravity tunnels with no exit, *Am. J. Phys.* **86**, 924 (2018).

[400] L. W. Taylor, *Manual of Advanced Undergraduate Experiments in Physics*, Addison–Wesley, London, UK (1959).

[401] I. H. Thompson and W. J. Whitrow, Time–Dependent Internal Solutions for Spherically Symmetrical Bodies in General Relativity: I. Adiabatic Collapse, *Mon. Not. Roy. Astr. Soc.* **136**, 207–217 (1967).

[402] I. H. Thompson and W. J. Whitrow, Time–Dependent Internal Solutions for Spherically Symmetrical Bodies in General Relativity: II. Adiabatic Radial Motions of Uniformly Dense Spheres, *Mon. Not. Roy. Astr. Soc.* **139**, 499–513 (1968).

[403] P. W. Thorp, Surface profiles and basal shear stresses of outlet glaciers from a late–glacial mountain ice field in Western Scotland, *J. Glaciol.* **37**, 77–88 (1991).

[404] R. C. Tolman, *Relativity, Thermodynamics and Cosmology*, Oxford University Press, Oxford, UK (1934), pp. 245–247.

[405] T. Torres, S. Patrick, A. Coutant, M. Richartz, E. W. Tedford, and S. Weinfurtner, Observation of superradiance in a vortex flow, *Nature Phys.* **13**, 833 (2017).

[406] M. Tsamparlis and A. Paliathanasis, Symmetries of differential equations in cosmology, *Symmetry* **10**, 233 (2018).

[407] W. G. Unruh, Experimental Black–Hole Evaporation?, *Phys. Rev. Lett.* **46**, 1351–1353 (1981).

[408] W. G. Unruh, Sonic analog of black holes and the effects of high frequencies on black hole evaporation, *Phys. Rev. D* **51**, 2827–2838 (1995).

[409] W. G. Unruh and R. Schützhold, On slow light as a black hole analogue, *Phys. Rev. D* **68**, 024008 (2003).

[410] L. A. Ureña–López, Unveiling the dynamics of the universe, arXiv:physics/0609181.

[411] T. Utsu, Statistical study on the occurrence of aftershocks, *Geophys. Mag.* **30**, 521–605 (1961).

[412] T. Utsu, Y. Ogata, and R. S. Matsu'ura, The centenary of the Omori formula for a decay law of aftershock activity, *J. Phys. Earth* **43**, 1–33 (1995).

[413] P. C. Vaidya, Nonstatic Analogs of Schwarzschild's Interior Solution in General Relativity, *Phys. Rev.* **174**, 1615 (1968).

[414] P. C. Vaidya, The External Field of a Radiating Star in General Relativity, *Current Science* **12**, 183 (1943). Reprinted in *Gen. Relativ. Gravit.* **31**, 11–1209 (1999).

[415] G. Venezian, Terrestrial brachistochrone, *Am. J. Phys.* **4**, 701 (1966).

[416] E. P. Verlinde, On the origin of gravity and the laws of Newton, *J. High Energy Phys.* **4**, 29 (2011).

[417] E. P. Verlinde, Emergent Gravity and the Dark Universe, *SciPost Phys.* **2**, no. 3, 016 (2017).

[418] S. S. Vialov, Regularization of glacial shields movement and the theory of plastic viscous flow, in IAHS Publ. 47 (Symposium at Chamonix 1958, Physics of the Movement of the Ice), IAHS Press, Wallingford, UK (1958).

[419] H. S. Vieira and V. B. Bezerra, Quantum Newtonian cosmology and the biconfluent Heun functions, *J. Math. Phys.* **56**, 092501 (2015).

[420] H. S. Vieira, V. B. Bezerra, C. R. Muniz, and M. S. Cunha, Some exact results on quantum Newtonian cosmology, *J. Math. Phys.* **60**, 102301 (2019).

[421] A. Vilenkin and E. P. S. Shellard, *Cosmic Strings and Other Topological Defects*, Cambridge University Press, Cambridge, UK (1994).

[422] K. S. Virbhadra, Janis–Newman–Winicour and Wyman solutions are the same, *Int. J. Mod. Phys. A* **12**, 4831–4836 (1997).

[423] M. Visser, Acoustic black holes: Horizons, ergospheres, and Hawking radiation, *Class. Quantum Grav.* **15**, 1767–1791 (1998).

[424] M. Visser, The Kiselev black hole is neither perfect fluid, nor is it quintessence, *Class. Quantum Grav.* **37**, 045001 (2020).

[425] M. Visser and S. E. C. Weinfurtner, Vortex analogue for the equatorial geometry of the Kerr black hole, *Class. Quantum Grav.* **22**, 2493–2510 (2004).

[426] A. Voellmy, On the destructive force of avalanches, Alta Avalanche Study Center, USDA, Forest Service, Utah (1964).

[427] M. Vollmer, The freezing of lakes in winter, *Eur. J. Phys.* **40**, 035101 (2019).

[428] M. Vollmer and S. D. Gedzelman, Colours of the Sun and Moon: the role of the optical air mass, *Eur. J. Phys.* **27**, 299 (2006).

[429] G. E. Volovik, *The Universe in a Helium Droplet*, Clarendon Press, Oxford, UK (2003).

[430] G. E. Volovik, Induced gravity in superfluid ^{3}He, *J. Low Temp. Phys.* **113**, 667–680 (1997).

[431] G. E. Volovik, Links between gravity and dynamics of quantum liquids, *Gravit. Cosmol.* **6**, 187–203 (2000).

[432] G. E. Volovik, Superfluid analogies of cosmological phenomena, *Phys. Rept.* **351**, 195–348 (2001).

[433] G. E. Volovik, Effective gravity and quantum vacuum in superfluids, in *Artificial Black Holes*, edited by M. Novello, M. Visser, and G. Volovik, World Scientific, Singapore (2002), pp. 127–177.

[434] G. E. Volovik, Black–hole horizon and metric singularity at the brane separating two sliding superfluids, *J. Exp. Theor. Phys. Lett.* **76**, 296–300 (2002).

[435] R. M. Wald, *General Relativity*, Chicago University Press, Chicago (1984).

[436] P. Wang and R. A. Davis Jr., A beach profile model for a barred coast—case study from Sand Key, West–Central Florida, *J. Coastal Res.* **14**, 981–991 (1998).

[437] E. W. Washburn, The dynamics of capillary flow, *Phys. Rev.* **17**, 273 (1921).

[438] H. J. Weber and G. B. Arfken, *Essential Mathematical Methods for Physicists*, Elsevier/Academic Press, Amsterdam (2004).

[439] J. Weertman, Stability of ice–age ice sheets, *J. Geophys. Res.* **66**, 3783–3792 (1961).

[440] S. Weinberg, *Gravitation and Cosmology*. Wiley, New York (1972).

[441] R. M. Weiner, *Analogies in Physics and Life*, World Scientific, Singapore (2008).

[442] S. Weinfurtner, E. W. Tedford, M. C. J. Penrice, W. G. Unruh, and G. A. Lawrence, Measurement of Stimulated Hawking Emission in An Analogue System, *Phys. Rev. Lett.* **106**, 021302 (2011).

[443] C. M. Will, *Theory and Experiment in Gravitational Physics*, Cambridge University Press, Cambridge, UK (1993).

[444] C. M. Will, The Confrontation Between General Relativity and Experiment, *Living Reviews in Relativity* **4**, 4 (2001).

[445] L. D. Wright and A. D. Short, Morphodynamic variability of surf zones and beaches: A synthesis, *Marine Geol.* **56**, 93–118 (1984).

[446] M. Wyman, Static spherically symmetric scalar fields in general relativity, *Phys. Rev. D* **24**, 839 (1981).

[447] H. Xie, J. W. Zhao, H. W. Zhou, S. H. Ren, and R. X. Zhang, Secondary utilizations and perspectives of mined underground space, *Tunnelling and Underground Space Technology* **96**, 103129 (2020).

[448] S. H. Yang and Y. L. Shi, Three–dimensional numerical simulation of glacial trough forming process, *Science China (Earth Sciences)* **58**, 1656–1668 (2015).

[449] A. V. Yurov and V. A. Yurov, Friedman versus Abel equations: A connection unraveled, *J. Math. Phys.* **51**, 082503 (2010).

[450] A. Zampeli, T. Pailas, T. A. Terzis, and T. Christodoulakis, Conditional symmetries in axisymmetric quantum cosmologies with scalar fields and the fate of the classical singularities, *J. Cosmol. Astropart. Phys.* **05**, 066 (2016).

[451] D. V. Zenkov (editor), *The Inverse Problem of the Calculus of Variations: Local and Global Theory*, Atlantis Press, Paris, France (2015).

Index

Lightning Source UK Ltd.
Milton Keynes UK
UKHW020615190822
407538UK00001B/22

9 781800 613423